INTRODUCTION

Since its introduction in 1948 when the Land Rover was quickly adopted by the War Department as a utility truck, it has become an identifying symbol of the British military. From the jungles of South America to the frozen tundra of NATO's northern borders, this light utility truck has proved itself time and again. Indeed, such is its prowess and adaptability in the field it has been adopted by allies and former adversaries alike as the go-to truck. From firefighting to special forces transport the Land Rover is the everyman of the 4x4 world.

Its genesis was in a world of post-war austerity, and it was this that made the Land Rover the ideal military platform. It was made from straightforward engineering and had a utilitarian and fuss-free appearance with wonderful adaptability, all without too much damage to performance or function. Its steel ladder chassis, aluminium panels and leaf-sprung construction marked this new vehicle out as something that was field-ready.

First ordered by the British Army in 1949, the Land Rover was soon deployed to Korea with the British contingent. This baptism of fire cemented the Land Rover's reputation and what was initially intended as a temporary model for Rover proved its ability to tackle anything thrown at it.

British Army vehicles in Croatia, 1995. Note the Bedford truck in the foreground carrying a Land Rover Defender 90. (Sergeant Brian Gavin, US Department of Defense)

Detail of the rear light of an early 1949 Series I 80in showing the now globally recognized Land Rover badge. (Mark Griffiths)

Series III, FFR, 1st Armoured Division, Operation Granby. Occasionally the environment got the better of the Land Rover, and it was all hands on deck. Note the extra stowage on the bonnet and hard-top sides and the discarded sand track in the background. (US Army)

In 1956 the Land Rover secured its dominion in the British Army over the Austin Champ by being declared the main General Service (GS) truck. Available as either short wheelbase (SWB) or long wheelbase (LWB), the Land Rover took all in its stride. By the 1970s the Land Rover had been joined by two uniquely military versions: the air-portable ½-ton Lightweight and the snub-nosed One-Tonne 101 Forward Control 105mm Light Gun tractor.

Soon military Land Rovers were in use in a wide range of roles supporting troops and mandates worldwide. From snaking exercise convoys complete with mandatory Sankey ¾-ton trailers filled with stores bobbing along behind them, to guiding aircraft carriers through the Panama Canal, there was no stopping Solihull's finest. By the mid-eighties the leaf-sprung Series Land Rovers were slowly being replaced by the new coil-spring Defender, but many Series Land Rovers soldiered on well into the late 1990s.

In terms of comfort and economy the Defender was a quantum leap forward. Again, it was easily adaptable, with a huge range of versions available in three main-wheel chassis lengths of 90, 110 and 127/130in. The development of the Land Rover culminated in the Wolf Defender; with its upgraded engine and strengthened chassis, it proved itself a worthy platform for the Weapons Mounted Installation Kit (WMIK).

The 127 and 130 versions were often converted into ambulances by Marshalls of Cambridge or worked as tractor units for the RAF Regiment and Royal Artillery Rapier Squadrons. The 110s became the Truck Utility Medium (TUM) and the 90s Truck Utility Light (TUL). These were also produced as extra duty (XD) and heavy duty (HD) platforms.

By the end of the twentieth century some 140 nations were using the Land Rover in

A CAV 100 Snatch. 3 Royal Anglian on patrol, Afghanistan. Note the thermal insulation blankets, IED countermeasures on the roof and 'Keep Back' sign.

Series III Lightweights of the Honourable Artillery Company (HAC), London. A commander's vehicle with its unique pennant and wheel finish, with white detailing and a deep gloss finish.

A beautifully refurbished SWB Series II RAF Land Rover with a Spitfire at an RAF100 event. (Ben Skipper)

a range of roles and many former military models found their way into civilian use. These have included ambulances working with a range of humanitarian agencies and Snatch versions being bought by news broadcasters to protect their mobile news teams in war zones.

For the modeller there really is an eclectic range of models available, which will be covered later in the book. Whilst the full range of Land Rovers is vast, this title will look at those Land Rovers that are catered for by model manufacturers worldwide. From these kits there is the opportunity to model Cold War warriors going about their business in Northern European Harrier Hides, to Wolf WMIKs dashing from one firefight to the next.

As with all things there are a great many local variations, from the famous SAS 'Pink Panthers' to the RAF's mountain rescue teams, and no two are the same once they've made their way onto the Motor Transport Officer's books. This gives the modeller free reign over how to finish their chosen model.

This title will look at both the leaf- and coil-sprung versions as well as the Lightweight and Forward Control models. It will also look at some of the specialist vehicles that may interest the modeller and whet the appetite for some great diorama and conversion options.

Today, seventy years after its unveiling at the Amsterdam Motor Show in 1948, it's reassuring to know that the Maurice Wilkes design remains in worldwide use today. That some 70 percent of the original two million Series and Defender Land Rovers are still in use today prove that the Land Rover really is 'the best 4x4xfar'.

Design & Development

Post-war Britain saw surplus military Jeeps being bought up by farmers as a break from the horse or tractor for those lucky enough to be able to afford one. As with all things use was often enthusiastic and the already war-weary Jeeps were showing signs of wearing out. On his farm in Anglesey, Rover's Chief Engineer, Maurice Wilkes, found himself contemplating such a calamity before his light-bulb moment. Rather than buy a replacement from Willys-Overland why not produce his own? The model would reflect the Jeep's own utilitarian appearance and engineering as well as fulfilling a much-needed gap in the Rover range while adding to the post-war coffers.

By mid-1947 work had begun on the Land Rover. The team that gathered at Rover's Solihull plant consisted of Robert Boyle, Arthur Goddard, Gordon Bashford and John Cullen, all led by Tom Barton. Bashford sourced two surplus Jeeps to be used for the chassis' basic dimensions and mechanical arrangements. Following a successful project review in September 1947, the team produced a prototype within three weeks.

The vehicle stood out primarily for its centrally mounted steering wheel, to negate the design of left- and right-hand-drive variants. The dashboard was furnished with the original Jeep dials and the wheels were of two-piece wartime type. The Jeep pedigree was clear as most of the prototype's running gear had been sourced from the surplus vehicles. Its 80in wheelbase wouldn't disappear until the Mark 3 Series I Land Rover entered service in 1954.

The prototype was powered by a 1,389cc in-line four-cylinder engine, which generated 44bhp, some 15bhp less than the Jeep's. This new engine transmitted its power to the Jeep's original two-speed Spicer Model 18 transfer case through the Rover 10's gearbox, which featured a freewheel feature. This mechanism acted as an automatic clutch, allowing the driver to change gear without depressing the clutch pedal, thus limiting the use of the manual clutch to purely starting from standstill or stopping operations.

The use of this type of gearbox was inspired given that the proposed initial market for the Land Rover was to be the country user. This was clearly considered as the main market as the design team

Left: Maurice Wilkes. (Land Rover, Middle East and North Africa – MENA)

Below: The prototype Land Rover, 1947. The Jeep heritage is clear, yet there is something distinctly Land Rover about its form, especially the front. The interesting driving arrangement would be short-lived.

Bottom: Early Series I engine bay. Note the large air cleaner and bronze maker's plates on the rear bulkhead. (Mark Griffiths)

had incorporated a swing-down tailgate to allow for the easy loading of any number of agricultural items.

It was decided to construct the body from the post-war glut of aluminium alloy as steel was still rationed. It also negated the need to design, build and use expensive pressing tools as aluminium could easily be shaped by hand. The detail elements of grab handles, various brackets and the windscreen elements were all 'borrowed' from the donor Jeeps.

The resulting vehicle, presented to the Board in October 1947, bore more than a passing resemblance to the Jeep. However, what the Board saw they liked and the go-ahead for fifty pre-production vehicles was given to the team, with an official launch at the 1948 Amsterdam Motor Show planned. The prototype had proved two things: firstly, the idea was sound and secondly, to continue to use Jeep parts for elements of future models was impracticable in the long run.

Olaf Poppe's simple yet strong Series III chassis design stayed with the Series Land Rovers throughout their production. (Lewis Collard)

1949 Series I 80in. This beautiful example of an early Land Rover is exactly how early military vehicles would have been received. Note front capstan winch and covers. (Mark Griffiths)

The main change on the pre-production vehicles was the abandonment of the centrally placed steering wheel and the return to the more conventionally placed left- and right-hand-drive arrangements. The American instrumentation was also replaced with dials from Smiths, which continued to supply Land Rover with its instrumentation until 1999.

In terms of construction the aluminium bodywork was retained, though remodelled to include externally hinged doors as opposed to cutaway side sections. These were slab-sided and fitted with removable tops that included sliding side windows. The single-piece windscreen glass was replaced with a split-screen style that would become the defining feature for Series Land Rovers.

The chassis, which retained its 80in length, was completely redesigned; for this task Bashford approached Rover's Chief Engineer Olaf Poppe to see what could be done. Poppe's solution was simple, yet effective: the new chassis was made up of a ladder structure, manufactured from four pieces of box-sectioned steel welded together. The whole assembly was then galvanized for protection from Britain's somewhat temperate environment. Such was the success of this design that it remained in use until the final SWB Series III left Solihull nearly forty years later in 1985.

The Rover 10's 1,389cc engine was to be replaced by a larger 1,595cc engine with an improved 50bhp power output. The choice was also made to produce all vehicles with permanent four-wheel drive. The power was transferred via a specially designed four-speed gearbox and two-speed transfer case combination. The freewheeling mechanism was retained to prevent drive-line stress when traversing hard surfaces, which given its proposed uses as a go-anywhere vehicle made perfect sense. Of note to the modeller these early Land Rovers had elements cast in phosphor bronze which included the front hubs, and the tops of the brake and clutch pedals.

By the end of 1947 twenty-five of the fifty planned pre-production models had been manufactured, of which a number were rear-wheel drive only. By the time of the Amsterdam Motor Show in April 1948 Rover had produced forty-eight of the fifty pre-productions models. Soon the road tests would be released in the British motoring press and interest in what was meant to be a short-gap production model would soar exponentially.

Whilst Wilkes's intention had been to produce a replacement for his farm Jeep, the military applications were as clear as daylight. On 23 March 1948 Rover's board was made aware that the newly independent Indian Army was interested in their latest venture. Whilst the interest was welcome, it must have also been concerning given that at that stage the project was still very much embargoed from the public.

The British War Office and Ministry of Supply were not far behind, and in the early summer of 1948 two pre-production vehicles, a right-hand drive (R30) and a left-hand drive (L29) version were loaned to the Fighting Vehicle Design Establishment (FVDE). These vehicles were tested at the FVDE evaluation site at Cherstey, Surrey. Here the vehicles would be put through their paces on a range of tracks and tasks to check for reliability and durability.

Ironically, at the time the War Office was developing its own Jeep replacement, the Nuffield Gutty. This light field car project featured a semi-stressed skin construction with independent torsion-bar suspension and ran on a flat four. This in turn morphed into the Wolseley Mudlark which ultimately became the ill-fated Austin Champ.

The Austin Champ, designed as a combat support vehicle, was a complex and expensive machine; its service was cut short by the cheaper and simpler Land Rover.

The Champ, introduced in 1952, was an outstanding off-road platform; however, this didn't stop its production run being cut short by some 4,000 units. It was expensive, mechanically complex and unpopular with users, due in part to its open tub design and mechanical eccentricities. By late 1966 the Champ had been entirely withdrawn from service, by which time the Land Rover had established its dominance as the British military's light utility vehicle.

The seeds of the Champs demise at the hands of the Land Rover were sown even before it had entered service. As its genesis was prolonged, and with the Army's Jeep fleet not getting any younger, a solution was sought. The FVDE's evaluation of Land Rovers R30 and L29 were favourable, with some improvements suggested. The result of the initial trial was that a further twenty vehicles were ordered in October 1948 and collected the following December.

At this stage Rover were adapting their design to meet military specifications. Changes included the fitting of a full-length canvas tilt in khaki and single-piece wheels. A Pyrene fire extinguisher was also fitted on the inner bulkhead and a War Department (WD) pattern towing hook and electric sockets were added to the rear cross member.

When ordered, the Land Rover was intended to temporarily fill the Champ's role until it was ready for delivery to front-line service units. The Land Rover would also be replacing the war-weary Jeep fleet, even those units that were still in store. It would be a considered a commercial (CL) military vehicle, whereas the Champ was considered a combat (CT) military vehicle.

In military terms the CL class vehicles were non-specialist vehicles, bought off-the-shelf direct from manufacturers, to fulfil administrative roles with the minimum of modification and work well away from the front lines. CT vehicles were specifically designed to operate on the front lines as part of battle groups engaging in action against hostile targets with armoured fighting vehicles (AFVs) and supporting the 'Teeth Arms' directly.

1949 Series I 80in front grille. Note the unpainted spare wheel boss. (Mark Griffiths)

By the summer of 1949 Rover was committed to supplying nearly 2,000 vehicles to meet the Army's CL needs. Such a large order placed an initial strain on the Solihull plant, which was also building Land Rovers to satisfy civilian orders. To get around this issue Rover built the order in batches, and standardized the use of Deep Bronze Green as the base colour for the finish of all its civil and military Land Rovers.

Whilst this delivery period was ongoing, a further two production vehicles were subjected to a second series of tests. These tests were to prove whether or not the modifications recommended at the conclusion of the first tests had been made and were satisfactory. The two Land Rovers were put through their paces and then subjected to review at 10,000 and 20,000 miles respectively whereupon their engines would be stripped and wear measured. Whilst engine wear was such that the first overhaul was suggested at the 20,000-mile point, there were other more pressing issues. The first Land Rover, aside from the water pump being less than reliable and having to be replaced with a modified design, only had a few minor issues. This showed that the modifications requested by the War Office and FVDE after the first trial in 1948 had been successful.

The second Land Rover initially fared less favourably. The rear axle failed with a gearing issue before completion of its 20,000-mile review. Whilst undoubtedly infuriating, this was the result of poor assembly at Solihull and not a material issue. Once the axle had been replaced, the second vehicle continued to be trialled, covering 4,500 cross-country miles as party of its 20,000-mile trial.

In terms of consumables the trials saw the replacement of one set of spark plugs, a gearbox selector spring, four bushes, a spring and tyres. Overall a great result for what would have been a fair, yet arduous trial, and also not bad considering both Land Rovers were off-the-shelf purchases. By the end of the 1950s military Land Rovers had also received their baptism of fire as part of the UN forces in Korea and the Malayan Emergency.

To further boost the Land Rover's military credentials the RAF made the first of three purchases in 1950. Many of these vehicles ended up as quick-response vehicles with station fire sections. Equipped with rudimentary firefighting and rescue gear, these vehicles would race ahead of the heavier Bedfords and Austins, providing immediate and invaluable first response.

The Land Rover's development didn't stall, and in 1951 Rover upgraded the 1,595cc engine to a 1,997cc unit that delivered an extra 2bhp. The upgraded Land Rovers became known as 'Truck, General Service (GS). ¼ ton, 4x4, Rover Mk 2'. Not only did the Mk 2 benefit from an upgraded engine, but the headlights were moved from behind the engine grille, to be placed slightly proud of the grille.

Whilst the late Series I engine bay was a cramped affair, everything was easily located. Note the large circular air filter housing. (Mark Griffiths)

The Land Rover was now making a great impression, especially with the RAF, which in 1952 ordered further units to be delivered in both RAF Blue and Bronze Green. These went to the emergency service branches of the RAF – the police and mountain rescue teams – with some also going to the RAF Regiment.

The Royal Navy (RN), obviously aware of the impact that the Land Rover was making, made their first order in 1951 for their own Navy Blue Land Rovers. The vehicles would be carried on surface fleet vessels to work as shore-based liaison vehicles. Three RN Land Rovers were painted orange for use as first-response firefighting tenders at Royal Naval Air Service (RNAS) stations.

As a result of comments from customers about the size of the rear cargo area, Rover extended the chassis to 86in in 1953. They also introduced the 107in long-wheel-base chassis, known as the Mk 4, with its 6ft-long cargo area. While the Mk 3s were little more than civilian vehicles, they came with a unique soft top with extra straps and buckles that allowed the sides to be rolled up. The side lights moved from the bulkhead to the front of the wings and recessed handles appeared in the main body of the doors. For the first time a fighting vehicle number was allocated to a Land Rover, and the Mk 3 become FV18001. The first Mk 3s were issued in late 1954 and in 1955, with the RAF in particular placing sizeable orders.

The Mk 4 Land Rover was the designation given to *both* the 107in and 109in LWB which were mechanically the same as the 86in Mk 3. However, the Army wasn't initially that impressed by the arrival of the Mk 4 as there was still a considerable stock of 1-ton vehicles, which were still utilized at unit level. In 1955 the Fighting Vehicles Research and Development Establishment (FVRDE) reviewed the station wagon variant. Their investigations must have been favourable as the Army ordered twenty LHD units for evaluation some seven months before the production lines had produced their first example.

Admittedly very austere-looking, the Mk 4, classed as FV18004, was able to carry seven personnel. The Mk 4s were specified as being fitted for wireless (FFW) vehicles and were spread around the globe, with versions going to Central Europe and the Middle East. The 107 LWB was ordered in small numbers by the RN and RAF in pick-up and station wagon versions but only used for special duties such as bomb disposal.

The RAF took delivery of sixty Mk 4 pick-ups in 1956 and began converting these into airfield crash rescue first-response vehicles. The RAF's mountain rescue teams also took delivery of their first ambulances converted by Bonallack & Sons. The RN ordered fifty predominantly station wagon-type Mk 4s, which replaced the SWB Mk 1s in the airfield crash rescue role.

Outwardly, the SWB Mk 5 looked no different to the Mk 3 Land Rover even though the chassis had been lengthened to 88in with the front member relocated. Continuing to use the FV number of FV18001, the Mk 5 had been stretched to enable the fitting of the new two-litre engine, the size of which would have fouled key control mechanisms including the front axle. This new length would become the standard for all future SWB Land Rovers thereafter. Other changes included inward-facing rear seats and grab handles located either side of the tailgate.

In 1957 the War Department standardized the Land Rover as its light utility vehicle and replaced its ageing car-based vehicle fleet, like for like. Eager to please, Rover took on the challenge as the changes would be straightforward; however, the extra work involved made the 4x2 Mk 4s more expensive than the standard 4x4s. Thankfully, these costs were outweighed by the benefits, including a 10 percent increase in fuel economy.

The Ministry of Supply's initial order was for 625 fitted with canvas tilts and twenty station wagon variants. Most ended up fulfilling administrative duties including recruitment, driver training and liaison tasks. A few found themselves as loan vehicles in Hong Kong to the RAF Police Dog Sections.

When Series I production ceased in 1958, approximately 15,000 Land Rovers had been delivered to all three branches of the armed forces. At this point the Land Rover also accounted for 40 percent of the ¼-ton vehicles held on charge by the Army alone.

Series II

The first decade had seen numerous refinements to the original design, as well as ten years listening to customer feedback and reviews. In April 1958 Rover launched the Series II. The most noticeable change was the appearance. The slab-sided body was replaced with a styled body, featuring a waist pinch that ran around the body from wing to tail and the track increased from 50in to 51.5in. The engine was now a 2,286cc in-line four-cylinder OHV model with 51bhp. Electrics were either 12v or 24v. The fuel tank filler had moved from under the seat to a more sensible position on the side of the vehicle.

New military specifications included FV pattern lighting, stacked sidelights and indicators and reflectors fixed below the rear lights. A convoy light was also fitted which allowed vehicles to travel in convoy at night without resorting to main lights. The rear face of the back differential was painted white with a light directed at it. The system worked remarkably well as long as light and face were kept clean.

The Series II was identified as the Mk 6 for the 88in SWB and Mk 7 for the 109in LWB. The Series II was only produced between 1958 and 1961, so numbers in service were low, and the unit amalgamations and personnel losses as a result of the 1957 Defence White Paper ensured numbers purchased by the three services would never be great.

Interestingly, the War Office continued purchasing the 4x2 versions, ordering 275 Mk 6s for the purpose. All were essentially soft-top civilian-fitted vehicles whose only

The final version of the Series I (1956 88in) featured a longer chassis, remodelled doors featuring recessed exterior door handles and air vents below the window. (Mark Griffths)

Early Series II ex-military Land Rover in Bronze Green with yellow bridge plates and unit codes on the wing fronts, featuring 'pusher' bumpers and a small sub-unit identity badge on the grille. (Mark Griffiths)

A military innovation: a clear view of the convoy light reflector though the oil leaking from the diff may be a hindrance; it shows the difficulties personnel faced in keeping things clean. (Ben Skipper)

Right: The 1959 Cuthbertson conversion was a unique method for traversing marshy ground and was used by the RAF at the shoreline-based bombing ranges. (Mark Griffiths)

The M40 Portee SWB Series I Land Rover was an awkward-looking, yet effective set-up that would continue to be used up until the early 1980s. (Nick Dowling)

military fittings were a Pyrene extinguisher, rear reflectors and WD-pattern towing hook and Warner towing socket. These would later be replaced by Hillman and Ford estates with some of the 4x2s finding their way into service with other government ministries.

Over 1,100 Mk 6s were delivered to the RAF during their period of production in both RAF Blue Grey and Bronze Green. As well as ordering petrol versions, diesels were delivered from mid-1960 to work in flammable areas. Some also went on to fulfil a range of roles including bomb disposal vehicles fitted with the ingenious Cuthbertson track system. This simple conversion was fitted to six vehicles that worked the shoreline of the RAF bombing ranges.

The RN only took a small number of Mk 6s, fifty-six in total with fourteen diesel variants, and one destined as a barracks-based domestic fire tender. The remaining vehicles were of crew-cab design and destined to join the fleet's capital ships for liaison duties.

The LWB Mk 7 was mechanically the same as the SWB 88in version with the obvious exceptions of chassis length and associated changes. The military was still considering what tasks the 109in, often supplied as a GS pick-up with full-length canvas tilt, could fulfil, the most obvious being the ambulance role.

The first ambulances were ordered in 1959 and built on the station wagon chassis. These were designed and built by Mickleover Transport of London as two-stretcher types, with the bodywork designed for fully mobility and ease of use. The low roof allowed easy loading into contemporary aircraft of the RAF's Transport Command while the rear portion of the body was angled to allow for ramp usage. Later versions featured roof-mounted canvas-covered stretcher carriers that could be accessed by simple buckles should the need arise.

Some Mk 7s were purchased in station wagon configuration for use as FFW vehicles while some made found themselves at Fort Halstead, Kent, for trials as a possible portee

Series II RAF ambulance. This wonderfully restored ambulance features a wealth of detail including the unique finish of RAF ambulances.

platform for the L6 Weapon of Magnesium, Battalion, Anti-Tank (WOMBAT).

The RAF then took delivery of station wagon-type Mk 7s and four crash-rescue ambulances. A majority – 75 – were the new Airfield Crash Rescue Tender (ACRT) type, equipped with basic first responder kits. The RN took delivery of 156 units, including twenty station wagons used to support RN Explosive Ordnance Disposal (EOD) teams. The remaining 136 units were pick-up types used for logistics support.

As a result of the 1957 Defence Review, the Land Rover found itself top dog in military circles. Gone were the venerable one-ton Austin and Humber GS cargo vehicles and in came the Land Rovers. Another boon for the Land Rover was the introduction of the Larkspur radio system. Larkspur was smaller than its predecessors, which meant that battalion-level communications, could be carried by Land Rover fitted-for-radio (FFR) models.

In 1961 the Series IIA was released with its 2,286cc diesel engine, which generated 62bhp. By1969 the Series IIA experienced a minor facelift as a result of changing vehicle lighting laws in Europe. The headlights were mounted on the front of the wings, where they were incorporated into a lighting cluster along with the indicators and sidelights. A new cross-shaped engine grille replaced the inverted T-shaped grille of previous models.

Series IIA Land Rovers were classed as Mk 8 (SWB) and Mk 9 (LWB). FFR vehicles featured the new 'pusher' bumpers, twin fuel tanks and a 24v electrical system. The Army continued to purchase the 2,286cc petrol engine, presumably for its 70bhp. Most of the Army's Mk 8 and Mk 9 Land Rovers were destined for the new strategic reserve light role and as support vehicles for British Army of the Rhine (BAOR) units.

This later Series IIA CL type shows off its military-spec canvas tilt with its unique sides. (Mark Griffiths)

Air Portable General Purpose Land Rover. The APGP idea was shelved once the Lightweight became fully available.

In its new role the Land Rover was now known as 'Truck, General Service'.

Initially the Mk 8 and Mk 9 Land Rovers were supplied with commercial differentials and half-shafts; these were replaced with strengthened items in the mid-1960s. The only vehicles to retain the commercial elements were the station wagons, now known as 'Car, Utility, ¾ ton, 4x4',

Series IIA, Aden, 1967. Note the red oxide primer visible in the wheel arch. (Brian Harrington-Spier)

purchased in relatively low numbers and often undertaking military attaché duties.

The changes in how the military motor pool was organized opened the door to a range of interesting variants and experiments, one particularly ambitious project being the Air Portable General Purpose (APGP). The APGP was essentially an everyman's answer to the needs of air mobile and Royal Marine units, but due to the arrival of the Lightweight, the APGP was quietly shelved.

Another experiment was the floating Land Rover equipped with oversized inflatable wheels that allowed it to traverse water, but little else. The wheels were designed by Nicolas Straussler who had designed the duplex drive fitted to armour for use during the D-Day landings.

The most important development for the Series IIA was that of the Vehicle Projection Kit (VPK). These remarkable kits were the brainchild of Major Reg Pearce, REME, as a response to increasingly sophisticated attacks by paramilitaries in the Northern Ireland troubles. A composite of glass fibre panels and trapped air, these kits were fitted to SWB, LWB and ambulance Land Rovers.

The RAF continued to build up its Land Rover fleet with diesel-powered Mk 8s and Mk 9s, opting for the soft-top with the Mk 8, and side-windowed hard tops for the Mk 9s, which were ideal for traversing a busy aircraft pan, in avoiding damage to aircraft. Large numbers of 24v vehicles were also purchased to utilize the alternator's 90-amp output to supply ground power to its helicopter fleet.

The RN also continued with their purchases, modifying their Mk 8's chassis to an 80in length to enable shipboard use. They also purchased elegant-looking yellow ambulances, able to carry up to six stretchers, and constructed by Herbert Lomas Ltd, for use at RNAS stations.

In 1967 all Land Rover electrical systems were changed to that of a negative earth system, in line with industry-wide changes. New vehicles also featured a newly designed black instrument panel, with a key start and four-cylinder versions were fitted with a Zenith carburettor. The SWB version was known as the Rover 10 and the LWB became the Rover 11; these were the last Series IIAs to enter military service.

The Rover 10 was purchased as a CL-type vehicle while the military awaited the arrival of the new Lightweights. The type owed more to the civil market than the military one, with certain features missing. These included the pusher-style bumpers, no oil cooler and the suspension was not of the drop-shackle type introduced with the Mk 8 and Mk 9 Land Rovers. A small batch of LWB Rover 11s were also supplied to various Army establishments including King's Troop, RHA.

The first batch of CL-type SWB Land Rovers was assigned to the Joint Service Liaison Organization (JSLO) in Hanover, Germany with the BAOR. These vehicles would work with the West and East German authorities, smoothing the way for vehicle movement between West Germany and Berlin. They also saw service with the British Frontier Service (BFS) whose area of responsibility was primarily the inter-German border. Although a small service, they performed the important role of defusing often tense situations between the indigenous border forces to ensure safe passage for cross-border traffic.

As well as the ambulance and WOMBAT portee versions, there were several interesting Rover 11 variants. A small fleet of 109 hard-top 24v FFRs were specially converted for use as electronic warfare vehicles. Operated by numerous squadrons in the Royal Corps of Signals (RSigs), these Rover 11s were known by a variety of names such as 'Van, DF Equipment, ¾ ton, 4x4'. These vehicles operated in Europe and in Hong Kong with the Royal Gurkha Signals and were used for a variety of purposes, including intelligence-gathering worldwide.

One interesting adaption was the anti-mine roll bar, a heavy-gauge double hoop that fitted over the top of the Land Rover. Initially fitted to Land Rovers serving in Aden in the 1960s, they were still in use in 1980 with the Commonwealth Ceasefire and Elections Monitoring Force in Zimbabwe. These Land Rovers featured armour plating for driver and front passenger, and blast deflector plates in place of the doors. The upgrade was doubted until a vehicle detonated a mine, which resulted in the driver only breaking his leg.

The Series III is equipped with an anti-mine roll bar and blast deflectors as part of the Commonwealth Ceasefire and Elections Monitoring Force operations in Zimbabwe. (British Army)

Series III

The Series III Land Rover, launched in 1971, was the result of changing global pressures that were exerting themselves on Land Rover, especially from Asian manufacturers. The new Land Rovers featured a revised transmission set-up. This included a heavy-duty clutch that linked to a new synchromesh gearbox (4F1R), a new rear axle, new braking system, and for the 109in models stronger half-shafts.

Left: A standard Series III LWB Land Rover marked out by its new style plastic grille, with soft top. (Lewis Collard)

Below: This example of an RAF bomb disposal Series III SWB Land Rover, with its familiar red wings, is a great example of local changes that often occurred to Land Rovers in service. Note searchlight, FFR antennas and reused ammo box attached to the front bumper. (Alf Van Beem)

Bottom: 1980 Series III FFR, note the colour of various details, including front mounted NATO towing eye. (Alf Van Beem)

Externally not too much changed: the wire grille was replaced by a more contemporary plastic type, while the lights remained on the front of the wings. Other changes included sturdier door hinges, while on the left-hand wing a heater inlet appeared. Internally, the controls were tidied up and relocated, so that they could be reached while wearing a seatbelt.

The first military deliveries were mainly LHD versions for Germany and were a mix of GS and FFR, predominantly LWB as a result of the Lightweight coming into service. It was also the first time that the military purchased vehicles with the six-cylinder 2.5l petrol engine. This engine could produce 86bhp and remained in production until 1978 when it was replaced by the V8, which in turn was only fitted to special vehicles.

This period of service was one of great change for military Land Rovers; the Series II-based Lightweight, introduced in 1968, negated the need for an 88in Series III, which did serve in small numbers. It was also a period where the forward control (FC) models were entering service. Until the new FC was fully in service, the LWB was also utilized as a gun tractor for the Royal Artillery. Upon full FC integration into the motor pool, the LWB Series IIIs returned to normal duties. The 109in Series III continued to serve until 1998 when the Ministry of Defence (MoD) finally moved over to a diesel-only fleet in line with the rest of NATO.

As well as continuing to purchase CL-class 109in Land Rovers, identifiable by their lack of 'pusher' bumpers, there

Military Police Series III LWB hard-top with black-over-sand camouflage and a wealth of extra stowage including a bumper-mounted ammunition box. (Alf Van Beem)

Dutch Army LWB Series III with the markings on bumpers and flanks, as well as the unique placement of the indicators on the wing tops. (Alf Van Beem)

were several interesting variants including the Carawagon Tactical Command Post. This unique-looking vehicle was designed and fitted out by Searle for use by brigade commanders in the field as a mobile command centre. Other variants included Marshalls bodied ambulances, CL-89 drone support vehicles and Mould VHF radio system repair vehicles, operated by RSigs.

The RAF and RN continued with their deliveries of LWB Series IIIs including both LHD and RHD models. Whilst many were of military specification, CL station wagon models were also purchased for liaison roles. The RAF continued to purchase their crash rescue ambulances, as well as more helicopter support vehicles.

Deliveries continued of the Truck Airfield Crash Rescue (TACR) and Mountain, Desert, Jungle Rescue (MDJR), with two versions of the latter, a command and an ambulance, being purchased. The RN followed the other services' purchase leads and took delivery of soft- and hard-top versions along with station wagons and ambulances.

Whilst the 88in SWB version was delivered to the Army in a range of configurations including soft-top and station wagon, it wasn't purchased in great numbers. The RAF used diesel SWB models for general duties around airfields; often hard-topped, these Land Rovers were fitted with a front towing eye. RAF vehicles often featured a roof-mounted amber oscillating light with yellow visibility stripes painted around the waist for ease of identification on active airfields.

The Royal Navy continued to use SWB for shore-based duties, which were carried either on deck or, in the case of aircraft carriers, stowed below decks. Others were incorporated into the Motor Transport pools. Those that found themselves aboard were beautifully appointed in a gloss Navy Blue finish, with white hard or soft tops and the ship's name in white lettering on the doors with the ship's crest mounted on the grille. The filler caps of RAF and RN diesel models were often painted yellow as a visual cue to remind users of the fuel type. These SWB Series IIIs were rarely fitted with the 'pusher' bumper system.

Lightweight Land Rovers – Series IIA and III

The Lightweight was designed in response to a 1964 specification that sought an air-portable SWB Land Rover. To achieve this, the standard Land Rover would have to be modified, with the main effort addressing width. The desired width was 60in, to allow two units to sit side by side in the Armstrong-Whitworth Argosy. By the time the Lightweight was in service the Argosy had been replaced by the C-130 Hercules, at which point width no longer mattered. The adjusted weight was important, especially in helicopter movement.

The specifications were exacting: for the 12v model the unladen weight was to be 1,136kg, while the 24v model was to be 1,409kg. The total payload, including driver, was to be 455kg and it should be able to pull a ½-ton trailer over a range of 300 miles. The War Office wanted the steering, engine, suspension and drive train to be the same as that of the standard Land Rover. This meant that the weight savings had to come from adapting the bodywork.

The design team of Mike Broadhead, later replaced by Bob Seagar, Project Manager, and Norman Busby, with assistance from the FVDRE and their successors, the Military Vehicles and Engineering Establishment (MVEE) had a prototype ready for test by 1965.

The main task was to address was track width, given the War Office's insistence that the drive train specifications be unchanged – it was clearly too wide, so a compromise had to reached. The width of the bodywork was reduced by designing a new bulkhead and replacing the curved side panels with Series I-style slab-sided panels. The wings were replaced with slim protrusions upon which were mounted side and indicator lights. Headlights were placed on the front grille for the Series IIA version, before being moved to modified wings for the later Series IIA and Series III production models. Even with all these changes it was still too wide.

A compromise was reached with the newly established Ministry of Defence regarding axle designs. Sense had prevailed and the Lightweight would be fitted with redesigned drive flanges and a narrower axle with shortened half-shafts. Further weight savings were made in the use of civilian springs and removal of the oil cooler. Narrower 6.00x16 tyres were fitted to single-piece rims. The bonnet was redesigned and was the only piece that required new tooling.

Next the team revised which parts of the superstructure and bodywork were necessary. The intention was for the Lightweight to be transported by air in a manner of ways: from underslung rotary load, ordinary air cargo or dispatched at height via a Medium Stressed Platform (MSP). To achieve this, all non-essential fittings were removed, leaving a skeleton vehicle capable of action on landing. To that end it was deemed that the front bumper, windscreen, doors' rear-side panels, seats, soft top and frame could all be removed. These items would follow the deployed unit by other means and be reattached. These changes meant the Lightweight achieved its desired weight.

The first prototypes were delivered for evaluation early in 1966. By mid-1966 a short-run contract for six vehicles was made for further evaluation, with these becoming 'Special Project' vehicles. These were supplied as both RDH and LHD, in 12v and 24v and the name Rover 1 was chosen to identify the new vehicles. The vehicles were now classed as ½-ton vehicles by a new system which identified total load-carrying capability rather than that of the load bed.

The Lightweight was designated as 'Truck, General Service, ½ ton, 4x4, Rover 1' and given the fighting vehicle number of FV18101. However, the new Lightweight was more costly than the 88in vehicle it was replacing, so MoD-wide adoption at this point was not an option.

In 1967 the first orders were placed: 92 for the Royal Marines (RM) and 1,000 for the Army, with priority given to the RM. Once the RM vehicles had been delivered, the 1,000 units for the Army found themselves at home with the air-portable brigades of the 3rd Division. The RAF and RN also received small numbers while Rover kept four for research and development purposes.

The Series IIA Lightweights were soft-top and powered by the 2.25l petrol

The Series III Lightweight was finished in Bronze Green, and featured the later-style wing-mounted headlights. Note the canvas straps for the bonnet spare tyre. (Mark Griffiths)

Finished in RAF Grey Blue, this semi-stripped Series III Lightweight is equipped for use as an FFR. (Alf Van Beem)

Ready for action, this Series III Lightweight is finished in white-over-black Arctic camouflage. (Alf Van Beem)

Fitted as an FFR, this Series III Lightweight is unique in not having the hard top fitted for added security like later Lightweight FFRs. (Alf Van Beem)

Finished in High-Visibility Yellow, this former RAF Series III Lightweight certainly cut a dash on an otherwise drab military airfield. (Mark Griffiths)

engine. The oil cooler and 6.50x16 tyres also made a return as the weight savings that had been initially planned were now irrelevant with the arrival of the C-130 and more powerful helicopters. It could be said these new aircraft removed the need for the Lightweight as they could carry original Series Land Rovers with ease. As a result, the Lightweight was rarely transported stripped down.

There were several variants, including a 24v FFR model known as the FV18102, which had a demountable radio bank that sat widthways across the cargo bay. Powered by a large battery box, these radio banks were quite weighty as they were designed to be transported separately and reattached on arrival at the drop zone.

The Series IIA Lightweights remained in production until 1972; thereafter, the Series III models were used. Some were retrofitted with hard tops as they replaced older 88in models, especially those in command roles. The hard-topped version was fitted to improve the physical security of the vehicle and its contents and station-wagon tops were occasionally fitted. Hard-top vehicles were fitted with either a conventional swing door or a split tailgate and hatch system, which was favoured by the RAF. Interestingly, the hard top was based on the 88in model, which meant the body would overhang the sides of the Lightweight.

A WOMBAT portee was operated by the RM, which had its windscreen removed with a barrel rest taking its place. Line layer versions were also supplied and some were equipped with the VPK for service in Northern Ireland. The RAF operated the Lightweight with the RAF Regiment and on airfields as a glider recovery vehicle.

Finished in Gloss Yellow with a khaki tilt, it was often equipped with an amber oscillating light with a transparent panel insert over the heads of the crew.

The Series III Lightweight was far more numerous in military use with over 15,000 made by the end of production in 1985; some 4,000 were bought by the Dutch, Jamaican and Omani militaries. Built in either 12v or 24v versions dependent on role, all models were delivered as soft-tops and nearly all built as petrol versions. There was the odd diesel model supplied to the RAF to work in flammable areas, and some finished in the RAF's blue-grey livery. The RN also took delivery of the Lightweight, with vehicles being used in the GS role.

Variants included a conversion that allowed Lightweights to traverse the peaty ground of the Falklands Islands. This conversion involved 15.5in-wide low-pressure tyres, extending the mudguards outwards, and rerouting the exhaust so it ran up the left-hand side of the cab. A steering damper and heavy-duty sump guard were also fitted. The final element of the conversion was to move the spare wheel mount from the top of the bonnet to a remodelled bumper. A supporting box was welded to it to keep the wheel in place, and a further wheel added to the rear tailgate.

The RAF had a helicopter-starting version which ran on 24v to start the new Puma helicopter. The reinforced hard top featured a platform with a collapsible rail to enable personnel to carry out serving and starting procedures. Given the Puma was a deployable asset, Lightweights were equipped with winterization heating packages. Made by C.J. Williams, they featured a large heater which was fed with coolant from the engine. This would then be ducted around the interior, which was insulated with rubber matting with external flaps attached to the windows and radiator grilles to retain the heat.

This RAF Red Arrows support Series III Lightweight certainly turned heads at an airshow. (Mark Griffiths)

The Lightweight also fulfilled some high-profile roles, including a Red Arrows liaison vehicle and a ceremonial vehicle with the Honourable Artillery Company. HAC vehicles were finished in a gloss Bronze Green finish, with chrome bumpers. All upholstery and the spare wheel cover were white along with axle end and wheel nuts. Each vehicle also had thc HΛC's unique coat of arms emblazoned on the doors.

Forward Control

The forward-control (FC) variant of the Land Rover had two incarnations. The first was a 1962 development of the existing Series II 109in chassis, the second a military-specific design. The first FCs saw the driver's controls merely shifted up and over the front axle to enable a larger load space behind.

Whilst the idea was sound, the execution was less so, and keeping the existing 2.25l petrol engine was perhaps not the best move for a vehicle designed for a significantly increased cargo payload. The idea was plagued by mechanical issues and

A wonderful study of the sturdy 101 Forward Control. Note the winch cable attached to the front tow hook. (GENI)

as a result the military reception of the idea was reserved. The FVRDE did purchase a single unit in 1966 for test purposes and the RAF purchased some as domestic fire tenders. The design changed several times, with wheelbase lengths of 112in and 120in being produced for trials, which proved too large for the Army.

The fire tenders were a hit and saw use with the Army and several overseas fire brigades. The fire tenders could be found in Fire Engine Red, Matt Green and Matt Black over Bronze Green finishes. Known commercially as the HCB-Angus, this fire tender was well equipped with pump, ladders, hoses and a large water tank as well as carrying four firefighters. Some civilian FCs did find their way into service with examples serving as RAF radar operations vehicles.

The idea of a Forward Control Land Rover was clearly an attractive proposition.

Royal Artillery 101 Forward Control FFR. Note the '58 Pattern webbing on the sides ready for action and to give passengers more room inside. (Alf Van Beem)

1976 101 Forward Control. Note the pioneer tool arrangement along the front. (Steve Glover)

So, when a solution was sought for an air-portable gun tractor for the new 105mm Light Gun, the idea was revisited as Series Land Rovers weren't strong enough for the task. The FVDRE had a very clear idea of what they wanted, and by 1968 they released the design brief to Rover who assigned Norman Busby to oversee the project. The new vehicle was to be of FC configuration, with a 101in wheelbase, and be of conventional truck appearance. It was also expected to be able to be configured to other roles including ambulance and signals vehicle. By 1971 all the prototypes were ready and available for testing.

Known as FV19000 series, 1-tonne, Forward Control Land Rover, the vehicles were trialled, including extreme environment testing in arctic and tropical environments; some also went to the School of Artillery for user testing. This period would also test the practicality of a specially designed trailer with a powered axle that ran off a PTO shaft on the FC's rear. Whilst a good idea on paper, in practical terms the powered axle trailers led to a range of dangerous issues, including jack-knifing. Thankfully this issue was discovered on an unrelated expedition, and the 105mm gun, which had been the intended recipient of the powered axle had the feature omitted.

The FC featured a new chassis and bodywork, a 3l petrol engine generating 110bhp powering a Range Rover gearbox complete with a permanent four-wheel transfer case. This set-up drove special heavy-duty Salisbury axles, featuring larger than normal half-shafts. The suspension was a combination of semi-elliptical multi-leaf springs and hydraulic shocks with an anti-roll bar at the front.

A Nokken capstan winch was also fitted and could be fed through apertures front and rear to retrieve swamped ordnance. One important detail that the new FC carried over from previous 109in models was the front wheel flange that allowed entry in and out of the cabin for driver and passenger.

As the FC was also intended to be air-portable, it could be stripped, to a minimum weight of 3,500kg. This allowed it be carried by Wessex helicopter and Bristol Britannia aircraft. It could also be mounted in an MSP with a 105mm gun for airborne drop.

Offered in both 12v and 24v versions, all FCs would leave Solihull with a standard soft-top canvas tilt; sometimes they would be converted into box body units by third parties, such as Marshalls of Cambridge who made the ambulance version. Other versions that were available included the Royal Artillery Command Post versions fitted with Field Artillery Computer Equipment (FACE) and the Royal Signals Vampire signals-intercept vehicle.

By the time production of the FC ceased in 1978 some 2,300 examples were in service and the last remaining versions were phased out of service in the late 1980s to be

Continued on page 49

Perentie 6x6, Australian Special Forces, Afghanistan 2005. Vehicles were finished in a tricolour camouflage scheme of light brown and olive drab over dark tan. However, this scheme is soon obliterated by dust abrasion and muck accumulation as well as being faded by the sun. Of note are the ammunition boxes on the running board behind the driving position and the camouflaged spare wheel covers. Camouflage netting would often be wrapped around the front crash bars, while black hessian sheeting would be attached and rolled up against the upper roll-cage. Numerous Bergens and bags were hung from the vehicle and the motorcycle was often omitted so more fuel and water could be carried on the rear platform. Unit badges were not applied as the 6x6 was only used by Special Forces in the Afghan Theatre.

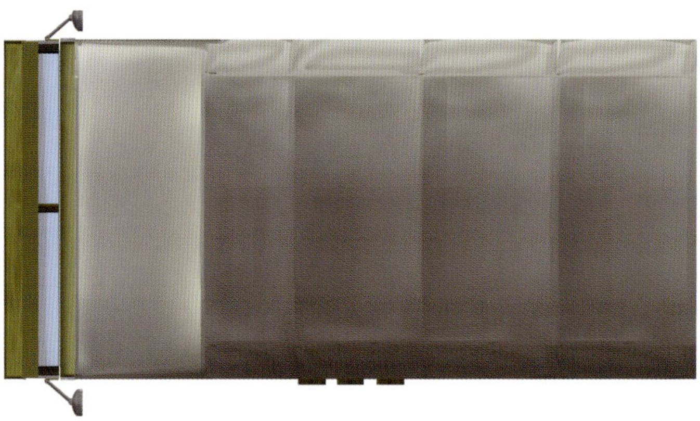

Forward Control FC101, British Army, Royal Artillery, 1975. Vehicles were either painted Bronze Green or Matt Black over bronze green; canvas tilt was khaki. When used, unit emblems were placed on the left-hand side of the front bumper or on the front bulkhead between the bumper and windscreen. They were applied on the right-hand-side panel of the rear panel next to the tailgate. The front wheel flanges would often lose paint on the raised tread areas. Early FC framework was either left unpainted or painted in aluminium. As all FC Land Rovers were petrol-powered, the filler caps were painted red.

Series IIA, Ambulance, Royal Air Force, 1965. Vehicles were painted RAF Blue Grey with yellow waist stripe, which ran along the sides inlaid with 'Royal Air Force Crash Rescue'. The large Red Cross symbol that was placed on the side covered these lines. RAF ambulances also featured a yellow roof and crash gate bars. RAF ambulances were always fitted with a centrally mounted blue light (in this case amber for transit); occasionally this would be complemented with lights placed between the gaps of the front bumpers. A small RAF roundel was often affixed to the front of the left wing and next to the bottom hinge of right-hand rear door. A light was positioned above the rear door on the left-hand side, angled downward to enable unhindered entry. Note that the extra stretcher carriers were not fitted on the roof as RAF ambulances were often fitted for four litters internally.

Defender 110 LWB GS Soft-Top, British Army, Royal Logistics Corps, 1995. Vehicles were finished in Bronze Green with dark green canvas tilt. Wheel arch spats were often left unpainted. These vehicles were heavily used and as a result there would be numerous little dents and scratches in the body work. Front wings often experienced the most damage, especially on the corners. The Bronze Green paint would fade rapidly once left out in the elements, the Matt Black, often applied locally and by brush, would rapidly turn to a 60 percent grey shade. Often the canvas tilts would be swapped around between vehicles, so had they been painted with Matt Black stripes, they would not match those of the vehicle. FFR 110s would often have the vehicle call sign chalked onto the doors.

Series IIA LWB, SAS 'Pink Panther', British Army, Oman, 1968. Vehicles were pink over Bronze Green; registration plate was carried front and rear; no other distinguishing markings were carried. These vehicles would have stores arranged that were mission-specific: more fuel, water and food for reconnaissance, more weaponry and ammunition for fighting patrols. Those prepped for use in temperate climates would retain their Bronze Green base coat. Camouflage netting and hessian would often be stored on the bonnet and along the sides and rear of the vehicle, sometimes draped over personal kit. Those painted pink had nearly every surface pained pink, from seats to tyre sidewalls; nothing was missed – even control pedals, jerry cans and fire extinguishers were painted pink. Those prepped for use in temperate climates would retain their bronze green base coat.

Series III Lightweight, Airfield Support Vehicle, Royal Air Force, Royal Air Force Central Gliding School, RAF Syserton, 1986. Vehicle is finished in a gloss yellow with bumper and support metal work in aluminium. Canvas tilt is khaki. Airfield support vehicles were always kept in immaculate condition. They were used for a variety of tasks, from bird scaring to glider retrieval. In the case of glider retrieval, a Perspex panel would be fitted in the roof or tilt above the driver to ensure they were clear to proceed. The retrieval rope was kept coiled in the rear of the cargo and an amber oscillating light would occasionally be fitted. Some glider squadrons placed a crest sticker either on the sides of the door, or on the front right-hand wing, with an RAF roundel sometimes also placed on the left-hand wing.

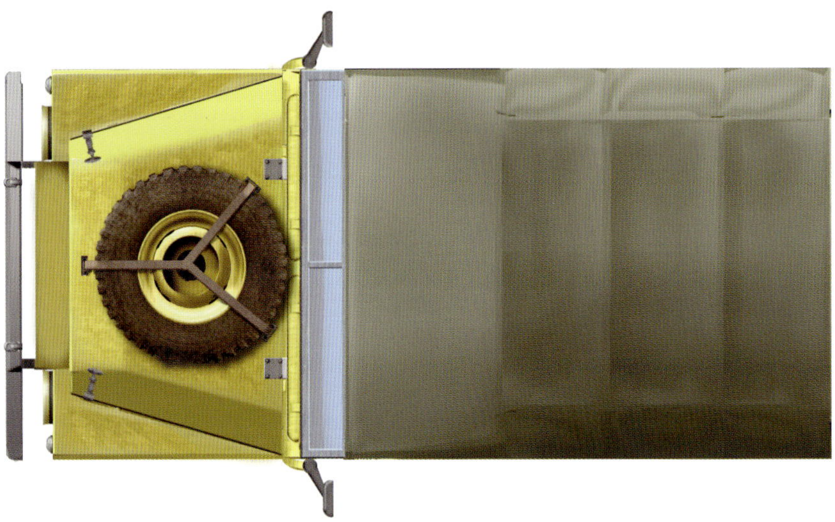

CAV100 Snatch Defender, British Army, 2 Para, Belfast, Northern Ireland, 1995. The vehicle is painted in Bronze Green. These vehicles were pooled and remained in theatre while Roulement battalions rotated. As such they were rarely adorned with any unit insignia. Occasionally a vehicle might have a small unit identifier badge on the front bumper and a discreet unit crest on the rear right-hand panel by the rear doors. Of note is the wire catchpole mounted on the left-hand side to snarl and break wires intended to harm troops in the observation hatch. Vehicles used to assist civil authorities were often the targets of petrol and paint bombs, so were sometimes seen with scorch marks or splashes of brightly coloured paint on their sides.

R- or E-WMIK Defender, British Army, A Squadron, Royal Yeomanry, 2015, Nottingham. Bronze Green, no unit markings. WMIKs in Yeomanry service are often kept a single colour with no Matt Black stripes, while medium grey seat covers are often fitted. Call sign indicators are placed on the side lockers located in front of the rear axle. As all Yeomany WMIKs are R or E versions, they are fitted with the later 15-spoke wheels; occasionally a canvas foul-weather tilt is fitted, but in practice a single Basha cover is simply attached to the front portion of the roll cage and affixed with bungees and paracord.

DEFENDER PERENTIE

Australian SASR Land Rover Defender Perentie, Afghanistan circa 2009

1/35 Scale

Brian Richardson

The Australian Defence Force has used the Land Rover series for many years and the Perentie (a large lizard of the Monitor family native to Australia) is based on the Defender 110 series introduced in 1987. Many versions were produced in 4x4 and 6x6 configurations by JRA Limited (now Land Rover Australia) for the Australian Army and part manufactured at Moorebank NSW in the 1980s and 1990s. This model represents an Australian SASR Land Rover Defender Perentie, Afghanistan circa 2009. The Hobbyboss Defender XD Wolf W.M.I.K. kit #82446 was converted with Black Dog's Land Rover Australian Special Forces big accessory set T35181. Hobbyboss's kit was released in 2010 and is just one of a series based on the same chassis. This is a well-moulded, quality kit that fits together well.

A few changes have been made by adding MRE boxes from Mouse House Enterprises as no two vehicles were ever the same and even the same Defender's load will change with mission requirements.

Black Dog's set is quite good, although the resin framework is brittle and needs a lot of straightening, filling and sanding. Facial tissue soaked in diluted PVA makes realistic camouflage wraps for the bull bar. Tamiya acrylics, Windsor Newton oils and Humbrol enamels took care of the painting stage and the few decals used on the ammo liners are made by Arms Corp Models ACM 35861.

LAND ROVER SERIES II

Special Air Service, Long Range Desert Patrol Vehicle (LRDPV)

1/35 Scale

Brian Richardson

The Special Air Service has made use of the LRDPV (Long Range Desert Patrol Vehicle) for many years and those used in desert operations were commonly painted in this distinctive colour, commonly known as Pink Panthers or Pinkies. Those operating in Europe were finished in the standard overall dark green and black. They served on operations in Malaya, Borneo, Aden and Oman where they were painted to meet local requirements. Tamiya released their Pink Panther in 1976 and it has been a very popular subject with modellers ever since. The kit represents a Series II long wheelbase model and includes a very comprehensive instruction sheet that has a detailed history including photos and drawings of Land Rovers' in SAS service, a very useful reference source. This was a feature of nearly all Tamiya kits at that time and is sadly missing from most kits today. This is my first build of this kit despite having it in the stash for many years and it turned out to be a fun build.

10 FG 63

The few extras that went into this job were a set of Mouse House bar tread tyres (MH 115), an FN SLR, also a Mouse House (MAW 704), and Live Resin's M3 Carl Gustaf recoilless rifle (LRE-35326) to dress it up. The stowage is scratch built or comes from the spares box. The search light is moulded solid so the lens was drilled out and filled with two-part clear epoxy after it was painted silver. As luck would have it, an air bubble was trapped in the setting epoxy and looks very much like a bulb.

All wood parts were painted with burnt umber over tan acrylic and the grain created with a stiff flat brush. Turn indicators were painted Tamiya X26 Clear Orange over Humbrol silver.

The rear number plate was made from Verlinden dry decals rubbed down onto a piece of decal paper painted black first.

LAND ROVER 109' LWB

General Service Vehicle, IFOR, Bosnia, 1995–1996

1/35 Scale

Brian Richardson

Italeri's Land Rover has been around since 1980 and has been re-released over the years in many different versions with this boxing coming out in 2012. This is built basically out of the box with just a few improvements. Left over Hobbyboss vinyl tyres from their WOLF kit were grafted onto Italeri's wheel hubs after removing the moulded-on tyres which have very little tread. The pattern isn't quite correct but is better than no tread at all which lets this otherwise good kit down.

Italeri's window parts fit well but aren't as clear as they could be. They've been used in any case and the side glass has been scribed to represent the two-part sliding feature.

Salami slices from thinly stretched sprue were added to each wheel nut, an easy fix for the missing wheel studs. A wire valve stem for each wheel was also fitted. The two tow points on the front bumper were added with copper wire; not all Land Rovers have them though. Tie straps for the spare wheel were added with Tamiya masking tape and copper wire loops.

Tamiya's XF65 Field Grey and XF69 Nato Black with X26 Clear Orange and X27 Clear Red over Humbrol silver were used for the side lights.

Italeri's IFOR decals are too light for olive drab but do respond well to decal softener when applied to the canvas cover. Italeri are way off with the colour of IFOR for their box art and decal sheet. Even their paint and decal guide is pretty generic with information.

SERIES IIA FIELD AMBULANCE

212 Field Hospital, RAMC, 49 Brigade, Sennybridge Training Area, Wales, 1995

1/35 Scale

Ben Skipper

This kit, launched in 1976 as a partner to the SAS Pink Panther kit, is an enjoyable and quick build and, in usual Tamiya style, well designed. However, it is ripe for a few improvements so the roof-mounted canvas-stretcher carriers were detailed using heads of track pins to stimulate press studs with thin strips of plastic for straps. The internal stretcher tie downs are made of foil and finished with restraining hooks and loops of fine wire. Blue lights have been added to the grille and the rear doors large hinges have been replaced with strips of bent wire that allows the doors to be removed for interior viewing. Finally, replacement Goodyear tyres from Accurate Armour were added and a small retainer for the bonnet spare wheel was fashioned from wire and metal strip.

Two three-quarter views: a left-hand-side front profile showing the extra width of the Marshalls body. Note the windscreen wipers and bonnet clips and a right-hand-side rear profile showing the slopping back of the ambulance body, designed not to foul the vehicle while ascending/descending a ramp.

The interior was painted first, with the driver finished in British Soldier 95 uniform. As British military ambulances rarely carry unit emblems the model was finished in temperate matt black over dark green with decals including Chalk Movement information. The safari roof was painted khaki and the all-important white convoy light reflector was added to the rear differential.

Full-length, left-hand-side showing light weathering, employed throughout. Black ink wash has been used on the wheel rims and a light dusting of tan was sprayed over the model.

Note the large press studs and straps made for the roof-mounted stretcher carriers. Note also the slightly modified driver figure in British Soldier 95 uniform.

A forward view of the left-hand side. Note the PSI reminder over the front wheel arch.

An overhead shot of the bonnet showing off the scratch-built spare wheel clamp and Accurate Armour wheel.

A lower three-quarter, right-hand rear profile showing off the rear portions of the chassis and exhaust pipe that was routed out thought the back of the rear wheel arch.

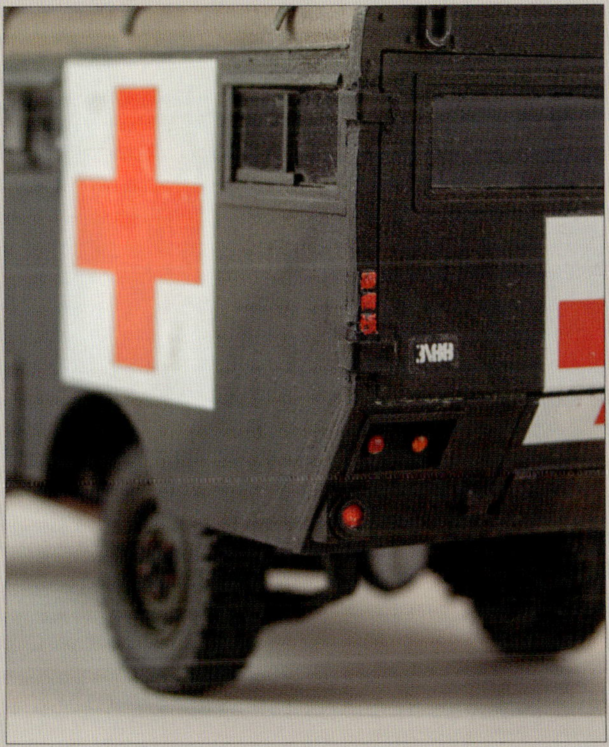

Rear light and reflector cluster. Note the lower circular reflector mount has been painted over to give it the same colour as the body.

Head-on view showing the detail of the Accurate Armour tyres and the grille-mounted blue lights.

A lower three-quarter, left-hand front profile shot showing off some of that wonderful drive-train and chassis detail.

Interior of the ambulance body showing off the scratch-built stretcher retaining straps as well as hook and loops. Just visible to the rear is the head of the white oxygen tank. Here the sliding glass panel to access the front cabin is visible.

The Land Rover has been in military service for over seventy years which is proof positive of the soundness of its original design and ensuing popularity. As a subject, the Land Rover is well served by model manufacturers, with a huge range focusing on key models in a range of scales and media. All kits have their merits with great potential for conversions as well as super detailing. They can also be incorporated into both small vignettes and large-scale dioramas with other genres of model making, including airfield scenes.

MODEL KITS OF LAND ROVERS

- Accurate Armour No. LR012 and LR013, 1/35, Series III Long Wheelbase Land Rover, Resin and Photo Etch
- Accurate Armour No. LR008, 1/35, Land Rover XD TUM Op' Telic Patrol, Resin and Photo Etch
- Accurate Armour No. LR009, 1/35, SAS Land Rover 110 DPV, Resin and Photo Etch
- Black Dog No. T35189, 1/35, Australian Perentie Land Rover 6x6 Complete Kit, Resin
- Black Dog No. T35191, 1/35, Australian Perentie Land Rover 6x6 Big Set Complete Kit, Resin
- BSM No. 350002, 1/35, Land Rover Santana S2 Ligero CSR 'Sahara Occidental' (Resin and Photo Etch)
- Firing Line, FL-MV-131a, 1/35, Land Rover 1-Tonne General Service (Forward Control), Resin, White Metal and Photo Etch
- Firing Line, FL-MV-132a, 1/35, Land Rover Lightweight Soft Top, Resin, White Metal and Photo Etch
- Firing Line, FL-MV-132b, 1/35, Land Rover Lightweight Hard Top, Resin, White Metal and Photo Etch
- Hobbyboss, No. 82446, 1/35, Land Rover Wolf WMIK, Plastic
- Hobbyboss, No. 82447, 1/35, Land Rover Wolf WMIK w/MILAN ATGM, Plastic
- Hobbyboss, No. 82448, 1/35, Land Rover Defender 110 Hard Top, Plastic.
- Hobbyboss, No. 82449, 1/35, Land Rover RSOV w/ Mk 19 Grenade Launcher, Plastic
- Hobbyboss, No. 82450, 1/35, Land Rover RSOV w/ MG, Plastic
- Italeri, No. 3644, 1/24, Range Rover Classic 4x4 Plastic
- Italeri, No. 3661, 1/24, Range Rover Two Door Police 4x4 Plastic
- Italeri, No. 6501, 1/35, Land Rover Series II 109 'Pink Panther', Plastic
- Italeri, No. 6508, 1/35, Land Rover Series II 109 Soft Top, Plastic
- Italeri, No. 6542, 1/35, Land Rover Series II 109 Guardia Civil, Safari Top, Plastic
- Kingfisher, 1/76 Land Rover individual models, Resin and White Metal kits
- Kit Form Services, TQ-204, 1/24, Truck 4x4 ½-Ton Airportable, Resin, White Metal and Etch
- Kit Form Services, TQ-204, 1/24, Truck 4x4 1-Ton Forward Control, Resin, White Metal and Etch
- Kit Form Services, TQ-228, 1/24, Truck 4x4 Series 2A Ambulance, Resin, White Metal and Etch
- Mil Mod Military Models, MM000-130, 1/72, 1-Ton Fire Appliance Land Rover Type (Royal Navy, Royal Bahrain AF, Belfast City Airport), Resin
- Mil Mod Military Models, MM000-130, 1/72 1-Ton Fire Tender, FC, FI Bates Land Rover Type, Resin
- Mil Mod Military Models, MM000-196, 1/72, Land Rover 3, ¼-Ton 4x4 Hard Top/Windows & Trailer RAF with Figures, and Load, Resin
- Mil Mod Military Models, MM000-197, 1/72, Land Rover 3, ¼-Ton 4x4 Hard Top/Windows & Trailer (Royal Navy, Marines) w/Figures, and Load, Resin
- Mil Mod Military Models, MM000-205, 1/72, Land Rover Series I 107-inch, Pilot Rescue (RAF Mountain Rescue), Resin
- Mil Mod Military Models, MM000-206, Land Rover Series I 107-inch, Pilot Rescue (Royal Navy, RAF), Resin
- Model Miniature, MM-R035, 1/72, Series III LWB w/106mm Gun, Resin
- Model Miniature, MM-R036, 1/72, Series III LWB w/DShK, Resin
- Model Miniature, MM-R192, 1/72, Defender 110, Hard Top, Resin
- Revell, No. 07047, 1/24, Land Rover Series III LWB, Safari Top, Plastic
- Revell, No. 03246, 1/35, Land Rover Series III LWB, Hard Top, Plastic
- Tamiya No. 35076, 1/35, SAS Land Rover Pink Panther Plastic

Accurate Armour

Accurate Armour's 1/35 resin and PE sets for the Land Rover, continue their tradition of producing excellent kits which provide the modeller with a range of subjects. All kits are supplied with a range of sharp markings for adding those finishing touches and instructions are always clear and well designed. The Series III LWB models cover the hard- and soft-top versions. The soft-top kit can easily fulfil a number of roles, from a company vehicle on exercise in northern Germany to a supporting Bosnia Implementation Force (IFOR) role. The hard top is available as two separate models: one, a solid-sided version or with windows, the other decked out as a helicopter support platform. All three kits come with a beautifully rendered 2.251 engine. The dashboard details are crisp and the rear sections for the hard- and soft-top version come with seating. It's the small details that really help these kits stand out and the NATO Dixon-Bate tow hooks, for example, are beautifully rendered. The vehicle's exteriors are supplied ready to accept the placing of a range of pioneer tools and

The LRPV is a typical of the well-observed models produced by Accurate Armour and is a wonderful subject for a Gulf War-themed collection.

This wonderful conversion kit produces an accurate model of 1990s CAV 100 Land Rovers deployed in Northern Ireland.

fitted-for-radio (FFR) antenna housings. The Land Rover XD WOLF Truck Utility Medium (TUM) can be fitted as the General Service (GS) version with soft-top canvas removed and in-theatre anti-riot screens and wire cutter added for early patrols in Iraq. The SAS Land Rover is a wonderful kit of the SAS's unique 110 Defender long-range patrol vehicle (LRPV). It can be modelled for either desert or temperate operations and is bristling with detail and equipment. This model really is the definitive SAS Land Rover and a great focal point for a Gulf War diorama. Conversion kits are also available to update Hobbyboss kits into either the CAV 100 or updates to E-WMIK or R-WMIK state. Again, all are wonderfully cast and the WMIK-based conversions have been made with the full assistance of Land Rover, as well as Ricardo Special Vehicles. For those just wanting to tweak their kits, Accurate Armour produces four types of wheel set. These fit all Land Rovers from the Series II to the last CAV 100s and R-WMIKs. These are beautifully sculpted and will enliven any version of the Land Rover you choose to make.

Black Dog

Black Dog's kits and updates offer the modeller options in both 1/72 and 1/35. Black Dog's castings and PE are as sharp as ever. They offer two 1/72 110 Defender kits in resin, with PE details, both hard top, one in station wagon guise. The kits feature detailed underbellies, with snorkels supplied for both. The plain hard-top version can be finished as an FFR or support vehicle with roof-mounted oscillating lights, while the station wagon features a PE roof rack and details. For the 1/35 modeller there is a series of resin and etch conversion kits, including the Australian Special Forces, CAV 100s – one is fitted with thermal insulation – and WMIK versions. These are based on the Hobbyboss range of Land Rover models and offer updates that place your model in a contemporary setting. One set features extra personnel baggage that gives the vehicle an extended patrol look. All three kits use the Hobbyboss WMIK model as the donor. There's also an update set for Tamiya's venerable SAS Pink Panther. This kit features a host of etch add-ons and resin casts that really do bring the model to life. From re-sculptured seats to the SLR rifles and packed kitbags, Black Dog's attention to detail steals the day. For US modellers Black Dog has created several detailing sets for the RSOV made by Hobbyboss. These sets include crew figures, which are lacking from the original model, as well as extra ammunition boxes, new seating and contemporary military items such as CamelBak rehydration systems. Their final offering is the unique Australian 6x6 Perentie kit, which is available either as the basic kit, complete with motorcycle or as an LRPV with full load. Both kits are filled with amazing details, including a wonderfully detailed 4l Isuzu turbo diesel engine. As with all Black Dog kits, this well-observed kit, alongside the 4x4 Australian Land Rover conversion kit, would make a wonderful diorama. A separate set of Australian wheels is also available.

RSOV PE and resin set. This set breathes life into the RSOV, providing endless modelling possibilities.

This multimedia kit of the Austrailian Perentie 6x6 is truly for the connoisseur, well researched and packed with features that can only be achieved through multimedia modelling.

BSM

BSM produces a 1/35 kit of the Santana Land Rover Series II Ligero CSR as a portee carrier for the US M40 recoilless rifle in use with the Spanish Army. This Spanish-produced SWB variant of the Land Rover, used in the Western Sahara in the late 1960s and early 1970s, is a wonderful mix of resin, PE and 3D printed parts, presenting modellers with a real challenge. The model's castings are crisp, with details well defined. The M40 is wonderfully reproduced. BSM also produce resin and PE upgrade kits for the Italeri/Revell 1/35 LWB Land Rovers. The resin set consists of seats for the front and rear with all seats well sculpted; these can also be modelled folded up. The PE requires some reworking of the original kits but the instructions make this straightforward. The detailing allows the vehicle to be modelled in a variety of poses, including tailgate down. The outcome is a wonderful-looking Series IIa 109 that makes a great stand-alone sample model.

Firing Line

Firing Line produces two very important different 1/35 versions of the Land Rover that have gained almost cult status among military-vehicle enthusiasts: the Forward Control and the Lightweight. Both kits are moulded in wonderfully heavy resin, with most of the key fixtures such as seating and chassis incorporated in the moulding. The PE, supplied predominantly for detailing and doors, is sharp and is slightly thicker than normal. White metal is used for wheel sets, drive train and selected accessories, including pioneer tools. The kits are straightforward in their assembly, but the plans do need to be studied thoroughly and carefully before assembly. Firing Line also produces a white metal detailing pack that can be readily used to update any 1/35 kits and features everything from jerry cans to steering wheels. There are also update kits for hard and soft versions of the Italeri/Revell kits featuring new wheels and body fittings. These include windscreen elements and PE for the FFR racking and radiator grilles.

This Series III Lightweight Land Rover depicts a vehicle serving with Air Traffic Control, Royal Air Force Central Gliding School, RAF Syserton, Nottinghamshire, 1983. This view, without tilt and frame, shows how utilitarian the Lightweight was.

Full-length left-hand side showing the heavy weathering. I used a special mix of brown and black inks to bring out the details if the wheels along with black and brown pastel dustings to simulate various stains.

Registration plates were occasionally painted when the original was damaged or lost and the replacement had yet to arrive. Note the red NATO tow hook and electrical hook up.

Head-on view showing the distinctively angular Lightweight bonnet and spare tyre strap.

Hobbyboss

Hobbyboss gives the modeller some wonderful 1/35 kits of the TUM version of the Defender, providing three versions to choose from: the 110 hard top, WMIK and RSOV. Presented with wonderfully dynamic and action-packed box art by Vincent Wai, the kits are supplied as sharp castings with up to five sprues and a single transparent sprue. Separate rubber tyres, detailed chassis, 300Tdi engines, drive trains and interiors are also supplied. The rubber tyres are representative of the Goodyear G90 tyres fitted to the British Land Rovers, while the RSOV is supplied with correct Michelin XL tyres. The WMIK and RSOV versions have frets filled with accurate mouldings of weapons mounts and weapons. Instructions are crisp and the painting guides well researched and designed, and clearly presented. Decals are in register and are beautifully rendered; however, the registration plates are erroneous for British military vehicles. The quality of these models is such that they form the base for a range of interesting conversion kits by other manufacturers and are the go-to kits for contemporary military Land Rovers.

1/35 Wolf XD WMIK: a great kit filled with detail and ready for patrol in the deserts of Iraq or the woodlands of Northern Europe.

1/35 RSOV: a kit that can be modelled to show its use by a range of countries, including the Czech Army.

Italeri

Italeri has produced three 1/35 kits of the Land Rover: the SAS Series IIa Pink Panther, LWB Series III station wagon and a soft-top Series III. Based around a common chassis mould, each kit's details vary from the original that was released in 1980. Box art is a mix of action and portrait, employing some interesting backgrounds, especially with the Pink Panther. The Pink Panther is moulded in pink plastic across four sprues and is a beautifully moulded representation of the Marshalls conversion. The kit includes two sprues of SAS-related stores and weapons as well as authentic Series IIa grille details and desert tyres. Lighting details are moulded into the relevant pieces and the tyres have been updated to feature the correct desert pattern. The station wagon, with its unique double-skinned roof, is the latest incarnation of the trio, featuring six khaki-coloured sprues. Also included are some wonderfully crisp transparencies, including separate headlight lenses and a set of painting masks. The instructions are clear and a painting guide allows the modeller to fulfil a range of versions of this kit. Decals, for five versions, are also included; these are sharp and in register, and feature registration plates for several countries. The LWB soft-top Series III is cast in four grey plastic sprues and one transparent, capturing the shape of the LWB. It includes sharply rendered decals for Series

IIIs serving in the 1980s and 1990s, with options for a medical support and IFOR vehicle supplied. The kit can be built stripped down as the doors are of two-part assembly like the real thing. Of note, this kit was re-boxed and previously released by both Testors and Revell. All three kits are well proportioned and ripe for conversion, with the station wagon in particular being a great candidate for a bomb disposal or airfield support vehicle. Italeri also produces two differing styles of 1/24 of the two-door classic Range Rover, a civilian and police-type vehicle. Originally made by Esci and presented in a box with photographs of the completed kit, these can easily be converted into either Brixmis or service police vehicles. Moulded in multi-coloured plastic across five sprues and featuring rubber tyres, the kits feature clear instructions allowing the modeller to produce either LHD or RHD versions. The decal sheet includes licence plates for civilian vehicles, including British and West German vehicles, which will aid modellers building Cold War Range Rovers.

Kingfisher

Kingfisher produces a wide range of 1/76 military Land Rovers from Series II onwards. Produced as multimedia resin and white metal kits, they are beautifully detailed for their size, and would make great additions to Cold War dioramas as well as wonderful stand-alone models in their own right.

Kit Form Services

Kit Form Services (KFS) produce a range of 1/24 mixed media kits, and have an enviable reputation for their work in the scale. Producing three beautiful models of the Land Rover – the Lightweight, FC 101 and Series IIa Ambulance – they all come with a huge amount of detail. These kits are for the experienced modeller and feature sharply moulded resin, including clear-cast lighting elements, white metal details and vac-formed components. All come with the necessary pioneer tools and equipment to help them come alive. The Lightweight is modelled as a soft top, and can be built with the tilt sides up or down. The driving compartment is faithfully reproduced and the engine is ripe with opportunity to be super detailed. Decals enable the modeller to produce five different versions, including Lightweights in Royal Marines and Parachute Regiment service, where it can be modelled in a 'stripped-down' appearance.The FC 101 comes as a soft top, with the tilt again being able to have its sides rolled up. The engine is semi-visible, but all the drive train elements are present. The driving compartment, along with the troop-carrying section, is beautifully laid out, complete with driving controls and seating. This kit comes with decals for five different versions, including those in RM and Parachute Regiment service. The final model is the ambulance, equipped with full engine, driving compartment and ancillary items. Decked out as one of the later four-stretcher carriers, this kit is a must not only for the diorama builder, but also the modelling historian. It features correct blankets and first-aid boxes for the period and orderlies' haversacks complete the picture. This later version of the ambulance has a few extra fittings in the main body so these are also included in the kit. This kit includes decals for three different versions as well as a blue resin-cast oscillating light for the roof. Kit Form Services also produce a range of detailing parts in resin, white metal and PE including Marston matting, jerry cans and ammunition boxes. Other accessories include toolboxes, spanners, hand tools, chequer plate, cab junk and snow chains.

This multimedia kit is as imposing as the real FC 101 and faithfully captures the bulk of the original.

Packed with details and beautifully finished, this wonderfully handsome kit really is one for the mantelpiece.

Model Miniature

Model Miniature kits are a series of 1/72 LWB Series III and Defender resin model kits. The kits of the Series III have a figure and load of either a 106mm anti-tank gun or DShK .50 cal heavy machine gun, which is mounted on the rear cargo bed. The Defender model is the station wagon version and is supplied with a French National Police figure. The kits themselves are moulded in resin as key components: chassis, body interior, bodyshell, tyres and detailing parts. The castings are crisp and kits are supplied with stickers as opposed to decals.

Mil Mod Military Models

Mil Mod Military Models are 1/72 vehicles designed for the aircraft modeller, seeking to supplement their model aircraft dioramas. The models include the F/6 FC fire tender and RAF mountain rescue Land Rover. Some kits come with multi pose figures, diorama items and are easily convertible; in fact, Mil Mod supply the modeller with great conversion ideas in the model packs. The kits cover an eclectic and important range of first responder vehicles and while not huge in size, they are wonderfully researched and well moulded. Decals are sharp, instructions clear, in full colour with a hint of humour.

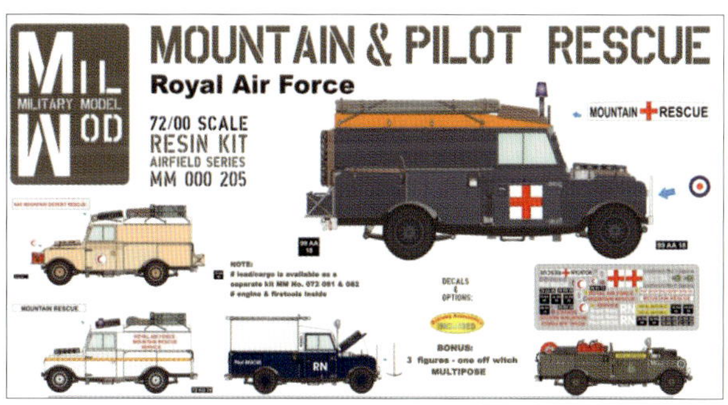

Mil Mod Models 1/72 resin model kits are quirky, fun and well presented with some of the best graphics on the market.

Revell

Revell's 1/24 Series III LWB station wagon, with safari roof, features a detailed engine and interior, and is supplied as an LHD version with a wonderful landscape scene, with the Land Rover off on its travels. Spread across seventeen sprues, moulded in a light grey plastic and including rubber tyres, this kit is full of potential, especially as a CL purchase for liaison duties on a Cold War NATO airbase. The instructions are the new style, easing the modeller through the build process. To finish, Revell has supplied a wonderfully sharp decal sheet for vehicles from a range of countries, as well as the all-important body and grille badges.

Tamiya

Tamiya's 1/35 Land Rover model is offered as the Marshalls' SAS Pink Panther which is presented in boxed form and featuring the wonderfully evocative artwork of Yoshiyuki Takani. The Pink Panther is cast in three pink sprues featuring realistic sand wheels, and a host of mission-critical equipment. Whilst possibly slightly out of scale, a fault of many Tamiya kits in the 1970s, it's still a great little kit for both the beginner and seasoned modeller. The instructions are clear and feature some wonderful reference illustrations and photographs of interest to the modeller.

DETAILING

There are numerous manufacturers offering additional detailing kits including decals. The one thing the Land Rover is not short of is etch and resin detailing kits, which cover a range of fits and configurations. The following items are available at time of writing, but this may change due to market conditions.

DEF Model

DEF Model has produced three packs of resin tyres that have been specifically designed for the Hobbyboss WMIK and RSOV versions. Set DW35007 is for later Defenders, such as the TUL and TUM and the Project Tithonus rebuilds, while set DW35008 is for later R-WMIK versions with their 15-spoke wheels. Set DW35027 is for the US Rangers' RSOVs and features the high-profile Firestone tyres. All tyres are weighted and moulded with ground contact bulges.

Djiti Model

This simple two-part soft-top conversion kit for the Hobbyboss Defender TUM hard-top kit features a replacement soft top and rear tailgate. The canvas tilt is beautifully finished, with exquisite detailing that shows the correct pull of the canvas over the frame beneath.

Dragonbadger Models

This firm supplies modellers with niche-interest 3D-printed conversions and kits. For the Land Rover modeller, they have produced a Cuthbertson track conversion kit for Land Rovers in 1/48, 1/35, 1/32 and 1/24 scale.

Eduard

Eduard produces two PE sets for the Land Rover. The first set, 35754, is for Tamiya's Pink Panther, upgrading or adding detail to the Tamiya kit. Spread across two frets and an acetate film, the PE requires the modeller to make some changes to the original kit; clear instructions help prevent overzealous plastic removal. The second set, 36193, is an ammunition box fret designed for use with the Hobbyboss WMIK series of kits. Straightforward to assemble, the kit comes as a single fret and illustrated instruction sheet. One interesting innovation is the incorporation of a series of hex shapes into which heated plastic rod can be inserted to make bolt heads.

ET Models
ET Models has produced three sets of replacement Land Rover tyres and PE sets for the WMIK versions. The tyre sets are for the Hobbyboss Land Rover and cover early ER35-020-, mid-life ER35-019- and late ER35-029-style wheels and tyres. Expertly moulded, the treads and wheel details are crisp and the wheels weighted to indicate ground pressure. The PE sets cover the basic WMIK version and the R-WMIK with the MILAN ATGM launcher. Both are well planned and highlight the differing versions. Set E35-178 focuses on the R-WMIK with the MILAN ATGM and features a wealth of detailing across six PE frets and one acetate sheet. Upgrade elements include netting, lashing points and canvas covers for the driving compartment. Set E35-109 features similar updates, including ammo boxes, across five frets and one acetate; it also features a resin .50 cal heavy machine gun for thc rcar gunncr position.

ET Models Hobbyboss 1/35 WMIK PE set.

Hussar
Hussar produces a set of resin wheels, HSR 35086, for Tamiya's Pink Panther model. The wheels are wonderfully rendered and capture Dunlop's sand tyre tread and side wall details accurately.

Hussar resin tyres for Tamiya 1/35 Land Rover.

Legend
Legend has produced four separate resin and mixed media sets for Tamiya's Pink Panther and Hobbyboss WMIK kits. These come in two very distinct styles. For the Pink Panther one set, LF1261, features resin-only elements; the other, LF1262, features resin and PE detailing. The details include kitbags as well as remodelled SLR carriers and Carl Gustav ammo containers. The PE edition features a single fret which includes a replacement radiator grille as well as sundials. Chains, brass rod and string for tow ropes are also included. The WMIK has two distinct sets. LF1226, which is marketed as the WMIK stowage set featuring resin baggage, seating and vehicle detailing parts and a single fret of PE for detailing purposes. This would suit a modeller seeking to portray a vehicle on operations in Afghanistan. Set LF1230 is a straightforward detailing set for early WMIKs and primarily features updated seating and weapons station mounts. This kit is supplied with a single fret of PE for detailing purposes. All kits come with well-illustrated and designed instruction sheets as well as decals for some items.

Minor
Minor has produced a workable PE spring set, VMD35022, for Tamiya's Land Rovers. The springs are fully functional and are built up in layers of PE that mimic true suspension. The instruction sheet is well laid out and keeps what seems like an intimidating task straightforward.

Suspension set for Italeri 1/35 Series Land Rovers.

PH Designs
PH Designs has produced a series of PE sets for the Hobbyboss Land Rover range. PH35LRG is a replacement front grille; PH35LRTP is a series of chequer plate cutouts for bonnet, wing and foot well. This kit would also be useful for producing Project Tithonus Defenders. The final kit is PH35WMIK1, which adds details to the driving compartment as well as wing and flank storage areas.

PHD's Hobbyboss 1/35WMIK etch set.

SKP Model

SKP Model has produced an accurate lens set (SKP127) for the Hobbyboss Land Rover range, which includes all lights, reflectors and wire headlight protectors. The lenses, a mix of PE and coloured resin, are beautifully crafted and in scale and could possibly be used for earlier Series models too.

Voyager

Voyager has several resin and PE kits that bring the Hobbyboss WMIK alive. The tyre sets, PEA264 and PEA277, feature early and later wheels respectively. These are well defined with excellent tread detailing with the later 15-spoke type really standing out. Their mixed media upgrade set, PE35431, features ten frets, one acetate sheet, a multipart resin and brass .50 cal and numerous rods. As with all such Voyager kits, this is breathtaking in its attention to detail and brings out the character of the WMIK.

DECALS

Star Decals

Star Decals have produced two completely different sets for military Land Rovers. The first, 35-C 1028, focuses on British and UN Land Rovers with UN, SFOR and IFOR markings. Beautifully rendered, a nice touch is the option to use hastily applied IFOR. As always, the blacks are 80 percent grey to ensure scale accuracy and the register of the Union flags and yellows are beyond reproach. Set 35-C 1239 features markings for a Series IIa Land Rover ambulance which served in the Croat Homeland War of 1991–95. The markings supplied include number plate, Red Cross and Croat Army shield. As always, the painting instructions are clear and give an idea of unit variants in theatre. The Croat markings are particularly interesting and reminiscent of late 1980s US Army-style camouflage.

Star Decals 1/35 Land Rover range; note the hand-painted IFOR decal, a great touch.

Kingfisher

Kingfisher produces a wide range of decals for the military Land Rover in 1/72–1/76, 1/48 and 1/35 scales. These include registration plates, NATO and UN mission markings, as well as staff cars, first responder and support vehicles. The decals are well researched and would match any number of roles you'd expect a military Land Rover to undertake.

KEY MODELLING ESSENTIALS

The military Land Rover was often as individual as its number plate, so its safe to say no two were the same once in service. Here are some helpful pointers to guide you towards making your authentic build:

- All Series-family Land Rovers are leaf sprung; Defenders are coil sprung.
- Series-family Land Rovers feature a flat split windscreen; Defender Land Rovers have a single-piece curved windscreen.
- Series I Land Rovers are flat sided, with no sills, and the headlights behind the grille.
- Series II Land Rovers feature a fitted sill and a waist-high pinch in the bodywork leading to curved outer edges.
- Later Series IIa headlights moved to the front of the wings.
- Series III have a plastic front grille, and air inlet situated on the side of the left wing.
- Lightweights have a jack locker in the left-hand rear-wheel arch.
- Defender front grille is flush with the wings.
- LWB Defender and Wolf models have internal stowage bins in front of the rear axle.
- Defenders feature rubber wheel-arch spats to compensate for wider tyres.
- Wolf-family Land Rovers have wheels on the side with bodies rising over the rear suspension.
- Pioneer tools are held by metal racks and canvas straps on the rear door, tailgate or top of the wings.
- On nearly all Land Rovers there is personal weapon storage behind or between the driver and passenger areas.
- Series engine blocks are normally Blue Green while later Defender and Wolf engines are black and silver.
- Paint is Black or Sand over varying shades of Bronze Green or Sand, often applied locally by brush.
- Paint schemes and markings change dependent on era and theatre modelled, e.g. NATO mandates.
- Diesel vehicles filler caps are yellow, petrol are red.
- Tyres and wheels are correct types for era and theatre, especially WMIK and CAV100.

replaced by the new 127/130 Defender and Steyr-Daimler-Puch Pinzgauer vehicles. Some FCs remained in service to see action in the first Gulf War and continued service with Territorial units until the mid-1990s.

A New Era: Coil-sprung Land Rovers

By 1978, the newly formed Land Rover Ltd found itself competing against capable Far Eastern 4x4s that were taking chunks out of Land Rover's market dominance. The Series models were showing their vintage. While their layout and build had become the stuff of legend, they were far from practical and the modern user expected a little bit more from a design and brand now over thirty years old.

Thanks to a sizeable government grant, Land Rover was able to develop new models, which resulted in the One Ten being unveiled at the 1983 Geneva Motor Show. The new One Ten, with its 109in chassis, was a quantum leap forward for both civilian and military users. The leaf springs were replaced by comfier coils and the split-plate windscreen replaced by a gentle curved single-piece unit. Seating was improved and the cabin layout more considered. Front discs were now in use and the grille had moved forward to be flush with the front of the wings.

Soon deliveries of the new One Ten were being made to the military, with most delivered as soft-top variants. There were key differences between the military and civilian One Tens The military versions retained their split-door panels and sliding windows as well as the 'pusher' bumpers. Door handles remained recessed and an externally accessed side storage compartment was added behind the cabin. Most were fitted with diesel engines in line with the NATO-wide diesel fuel policy.

Both GS and FFR versions were supplied, with FFRs having a hard top, and RAF One Tens supplied as station wagons. CL versions were also delivered, powered by either a diesel of petrol 2.5l engine, producing 68bhp and 83bhp respectively. The military used a range of variants, including Signals line layer and helicopter support types. A specially modified One Ten also replaced the SAS's famous 'Pink Panther' Land Rover.

The next model to appear was the One Two Seven Land Rover; a development of the existing 109in chassis that was simply stretched. Powered by the 3.5l petrol engine delivering 114bhp, the One Two Seven was available in a variety of configurations, including the crash rescue ambulance type. This particular adaptation was built by Marshalls of Cambridge and Locomotors of Andover. The Marshalls version featured a large window on the left-hand side, while the Locomotors version had a small manufacturer's plaque on the left-hand rear door. Other versions included the Rapier tractors which featured a specially designed tilt and replaced the FC 101s.

Royal Signals 101 Forward Control. Fitted with the radio van body, this 101 features a range of interesting details for the modeller. (Alf Van Beem)

101 Forward Control ambulance. Note the green diamond LOX (liquid oxygen) warning sign and completed bridge plate as well as the fold-up Red Cross plates on the flank.

Defender 90 hard top. The new coil-spring Defenders were welcome additions to any unit's motor transport fleet. Note how the original khaki paint has faded with time, which happened quickly in certain environments. (Mark Griffiths)

This Royal Logistics Corps EOD Defender 90 response vehicle lacks red wings, but has two oscillating lights to identify it as an emergency vehicle. Note RLC unit sticker on the wing and the bank of floodlights to assist low-light work. (Mark Griffiths)

Seen at the 2012 Sri Lankan Army Review, this Defender 110 pick-up features a custom-made VIP platform over the rear cargo area, and white-painted stripes on the tyres. (Chamal Pathirana)

The SWB version of the new Land Rover, the Ninety, wasn't available for use until 1986. With an increased wheelbase of 92.9in, it was capable of being powered by the same range of engines as the One Ten. It could be easily carried by the Puma and Chinook fleet as well as Hercules transport aircraft and because of this it made the lightweight series Land Rovers redundant. Delivered to all three branches of the armed forces, the Ninety, termed Truck Utility Light, was often used in the GS role and came equipped with soft-top, FFR hard-top and station wagon finishes; it was also purchased as a CL version.

The FFR version, which was fitted out for Clansman and later Bowman tactical communications systems, was used in a variety of signals roles at regiment and battalion level, serving the needs of the Signals Office, Quartermaster and Brigade liaison staff. The RAF also used the Ninety to fulfil a variety of airfield roles including Aircraft Armament Support Vehicle.

The Ninety found itself pressed into service with the SAS during the first Gulf War as a result of there not being enough One Ten vehicles available. These were arranged to match the layout of their larger versions, complete with weapons mounts, extra spare tyre and jerrycan side stowage. These smaller diesel-powered Land Rovers earned the nickname 'Dinky', though this was not in relation to their size but more of a play on words; a petrol-powered vehicle would be known as a 'Pinky'.

The Defender

By 1990 Land Rover were introducing changes to their range. The most noticeable was the introduction of the new 2.5l turbo diesel engine (200Tdi), which produced 107bhp. Other improvements included power steering, which was a boon as steering was heavy to say the least, and adjustable seating. The Defender name was chosen to reflect the Land Rover's military heritage with the name plate appearing on the bonnet as opposed to the grille.

Whilst the changes were welcomed, one aspect had been overlooked: the engine design could not accommodate the necessary 90-amp alternator that was vital to power the FFR's 24v electrical system. As a result, the military only purchased Defenders powered by the natural-aspired engine. After further development, the 300Tdi was released for use in 1994 which produced 111bhp and could house the required 90-amp alternator. As well as the new engine, these Defenders were fitted with a new five-speed gearbox.

Available as the Defender 90, 110 and 130, the vehicles were delivered as a 12v GS vehicle and 24v FFR vehicle. There were CL types for the RN and RAF which were easily distinguishable by their white roofs and gloss green bodywork. These were delivered as either station wagons or hard tops, and were often seen in RAF police or mountain rescue livery as well as special anti-spark vehicles which operated around armed aircraft.

One early variation was the specially designed Composite Armoured Vehicle (CAV) 100 Snatch Land Rover, designed by Courtaulds. This vehicle featured an armoured box which bolted onto the rear of a specially built chassis and replaced the previous VPK Land Rovers in Northern Ireland. Other variants of the 110 included a helicopter support and a winterized version for arctic warfare support, as well as SAS patrol vehicles.

The Defender 90 was not as prevalent as the 110, yet continued to fulfil the roles that SWB Land Rovers had fulfilled previously as 12v GS and 24v FFR vehicles. Supplied as both soft and hard top, the 90 was also purchased as the CL for use with the RAF on liaison duties. Some GS RAF 90s featured hard tops with side glazing to enable safe movement in aircraft handling areas.

The 90 could be carried as an underslung load by helicopter, stripped down for packing as an MSP load with another 90 for supporting airborne and air mobile assets and loaded on transport aircraft with ease. Like the 110 there were also winterized 90s working with both UK and NATO assets. These were identifiable by the Velcro strips fitted around windows to which the heat-retaining blankets were fitted while the heating system was in use.

The Land Rover 127 morphed into the Defender 130, and while this new name may hint at a larger chassis, it remained at 127in. The ambulance version, with its V8 engine, continued to be supplied while crew-cab pick-up versions were purchased for liaison purposes. A Mini Radar Van (MRV) was also produced for electronic warfare use.

Project Wolf

The Defender Wolf series entered service from 1996 onwards. While the shape was distinctly Land Rover, the vehicle was completely new and was the result of an MoD specification made in 1988. The new Land Rover would be available as 90, 110 and 130 vehicles. This new vehicle was developed in a changing world: the Land Rover's dominance in the Motor Transport sections was being challenged by the like of the Reynolds-Boughton RB44 4x4 and the Steyer-Pinz-Gauer 4x4.

This period was also one of great change: *glasnost* and *perestroika* saw a change in the Warsaw Pact that was irreversible, with the 'peace dividend' already being discussed in the US and UK. Armies would be smaller, funding cut and equipment costs challenged, so equipment had to be of the highest quality. Sadly, this was something that Land Rover had struggled with when the 200Tdi engine had been introduced, which meant the MoD had to make unforeseen purchases of V8-powered petrol versions.

New terminology was also used for the military fleet; 90s would become Truck, Utility, Light (TUL), 110s the Truck, Utility, Medium (TUM) and the TUM HD (Heavy Duty). This later model would be capable of carrying an increased load and towing heavy equipment. Land Rover approached Peter Marks, an army officer who had worked on the 127 Rapier and CAV 100 projects, to help the in-house military engineering team develop the new model. The first model, known as Challenger, owed more to contemporary Japanese automotive design practices than military practicalities. Based on the Discovery and called Project Juno, the exercise highlighted several issues, not least of all the new vehicle would struggle with military payloads, including the new NATO pallet.

As a standalone, Challenger was shelved by Land Rover's new owners British Aerospace, but the idea of putting forward a Land Rover for the new TUL and TUM contracts was very much alive. The Challenger had shown that Land Rover could still be innovative and had experienced engineers to make their bid a success. In November 1991 Land Rover was invited to bid for a lucrative contract of over 6,000 vehicles.

This repainted CAV 100 Snatch is a good example of former Afghanistan and Iraq vehicles without its thermal blankets and EOD countermeasures suite fitted. Note the various protective blanking plates and attachment points on the body (as circled). (Ben Skipper)

The last of the military Land Rovers, the Wolf TUL 90, finished in Bronze Green; this image shows how the body rises over the rear axle. (Mark Griffiths)

A Defender 110 TUM belonging to 16 Air Assault Brigade on patrol in Afghanistan, 2003. (MSgt Edward D. Kniery, USAF)

All vehicles were to be diesel powered and capable of operating close to the front line as part of a battle group, with certain vehicles conforming to a high-specification (HS) finish. A new Defender-based project called Wolf was launched to meet the MoD's requirements. It would utilize the existing range of Defenders, as well as develop

Note the Wolf's special sling attachment to raise and lower the flank-mounted spare tyre. (Mark Griffiths)

Honourable Artillery Company Defender TULs taking part in the Lord Mayor's Show, 2005. Note the unit crest on the door. (S. Pakhrin)

Now replaced on the front line by the next generation of wheeled vehicles, the R-WMIK is still in service with Yeomanry regiments in the British Army. (Stephen J Mason)

the 300Tdi engine. Other enhancements included disc brakes all around, new wheels designed and made by GKN, as well as the ability to take the recently introduced NATO cargo pallet, plus a higher rear roof section for ease of loading.

A total of eighteen prototype vehicles were delivered to the newly formed Defence Research Agency (DRA) for evaluation in 1993. Sadly, the testing period highlighted a range of deficiencies in the design that required rectification. The updated vehicles would become known as Wolf 2.

The cessation of trials also led Land Rover to rethink its own strategy and the TUM (HD) build was stopped, which allowed the military engineering team to focus on the task in hand. The new Wolf was redesigned and reinforced. The chassis received reinforcing cross-members as well as strengthening under the load bay. The suspension was upgraded and drive train elements strengthened. The body structure was also modified with a stronger bulkhead and sills, and the load bay reinforced. For safety a removable rear roll cage and anchor points for rear seatbelts were fitted.

Testing had also shown that the bonnet-mounted spare wheel could cause damage while traversing rough terrain, so it became side-mounted. As a safety measure, the wing mirror mounting arm was lengthened to allow unhindered vision past the side-mounted spare wheel. A new glass fibre hard top, with a pressing for a hatch, was also produced, which covered the roll cage and allowed its extra height.

The new vehicles were not officially known as Wolf but as Defender Extra Duty (XD). The MoD was suitably impressed with Land Rover's work and placed an order for both the 90 TUL XD and 110 TUM XD versions. The order would be a mix of LHD and RHD, fulfilling over twenty roles including GS and FFR, along with waterproofed and winterized versions. For airborne operations two TUL Land Rovers could be airdropped via MSP.

With the introduction of Bowman as the Clansman radio replacement, a series of five steps was made to allow for its installation from No Harness (NH) to Full Functional Local Area System (LAS) (FFL) internet connectivity-based communications. These vehicles would serve during some of the British military's busiest operational periods, fulfilling roles not envisaged by Wilkes at his farm in 1948.

In 2016, after some two million units had been produced, the last Defender rolled off the production line, capping almost seventy years of continuous work with militaries around the world. Thanks to various life extension programmes, such as Operation Tithonus, its ease of maintenance and, in many cases, lack of a suitable replacement, the Land Rover soldiers on, continuing to provide the modeller with a rich source of inspiration.

The Land Rover in Detail

For a design nearing seventy years old the Land Rover has changed very little from the first prototypes of the post-war years. Yet there is a wealth of detail waiting to be covered by modellers, especially when modelling the larger-scale offerings.

The Land Rover wasn't designed as a military vehicle from the outset so there are numerous similarities between the GS/FFR models and commercial variants. The Series I Land Rovers, aside from those specifically modified for special purposes, were extremely similar in appearance to the Land Rover doing the weekly market run.

This simplicity was the key to its success, and that helps the modeller greatly. The Series I vehicles were spartan: no sills and a steering column bereft of switches. The cockpit was fitted with a centrally placed instrument panel that gave the driver just enough information. Three passengers could be seated up front and the gear lever was cranked forward to allow the central passenger a little room. A small lamp was also provided on the passenger side to enable map-reading at night.

There was no external door handle, so a small canvas flap was built into the top half of the door. To gain entry you put your hand through the flap, and fumbled for the door handle, which can't have been much fun at night if you were cold, wet and tired. Later Series Is were supplied with a recessed door handle which made life easier and meant the cabin was reasonably weatherproof.

The spare wheel on the SWB Land Rovers, when not fitted to its bonnet boss, which often remained silver, would sit behind the cabin seats. The ribs on the floor of the small cargo bay would be either covered with a rubber mat or would occasionally remain unpainted.

The bronze finish of the front hubs would be painted over as would the metal-strengthening sections. Occasionally these would be exposed and polished if in use as a VIP vehicle. For RAF vehicles it was common practice to place a roundel on the front of the left-hand wing, and for RN use the ship's crest would sit on the front of the grille. RAF and RN soft-top tilts were issued as either Blue or Blue Grey.

Registration plates would be mounted on the front right and rear right wings. Bumpers were either flat aluminium or

Devil in the detail. Fuel caps were often colour-coded: red for petrol/gasoline and yellow for diesel. (Ben Skipper)

Utilitarian with centrally mounted gauges, the Series I interior was an exercise in post-war austerity. (Mark Griffiths)

Series I profile shot showing the canvas flap in the door that covered an aperture to enable access to the door handle. (Mark Griffths)

Right: Series II bonnet; the spare tyre attachment point and securing strap. (Mark Griffiths)

Below: The new inverted T-shaped grille reveals the headlights in this later Series I. Note the recessed door handles and unpainted front wheel hubs. (Mark Griffiths)

Desert Warrior: this Series I is finished in sand camouflage with a wealth of unit markings. (Les Chatfield)

Series IIA: note bridge plates and buckles on the side of the tilt for keeping the side in place once rolled up. (Mark Griffiths)

painted over; however, this would often flake off leaving patches of bare metal exposed. Of note was the reposition of front side lights and indicators, which had moved to the front of the wings by the time production ceased.

The Series II and IIA Land Rovers were fitted with a sill; early vehicles often had the metal of the windscreen and bumpers unpainted, especially for those classed as CL vehicles. RAF Land Rover roundels moved to the top of the right front wing in some circumstances and a registration plate was mounted centrally on the front bumper. On soft-top vehicles the sides of the tilt could be rolled up to enable ventilation and an unobstructed view.

The yellow bridge plate was now in common usage and was affixed to the right-hand side of the vehicle, either on the wing or grille. Unit markings also started to appear on the front wings. At the rear pioneer tools were often attached to the tailgate and the round, white-edged reflectors were fitted below the rear light cluster. On the left-hand rear wing just above the bumper were the connectors for the trailer lights, which would be painted either yellow or blue. The towing hook was painted black or red, and, on occasion, just the upper jaw was painted red.

Later Series IIs and IIAs in RAF service wore a high-visibility yellow band around the waist on vehicles that traversed active airfields. Bomb disposal vehicles in all three services were painted with red wings, front and rear. Internally the instrument panel was now painted black. There was still the capacity to fit three passengers in the front.

Series III Land Rovers were now fitted with a plastic grille that was often painted over. The front cabin was fitted with seatbelts and the instrument panel moved in the front of the driver. Lighting control dials were fitted centrally onto a black panel. The tow hook and electric hook-up for the trailer were now side by side, with the hook-up cover often painted red. Rear lighting arrangement also changed.

When the Defender Land Rover came out, the centre seat was omitted, occasionally replaced by a small locker. A new-style padded steering wheel was also added. An external fuel tank filler cap was fitted behind the right-hand door and on LWB vehicles symmetrical jerrycan lockers were fitted in front of the rear axle on the sides of the vehicle. CL versions were little changed from their civilian counterparts, and featured the comfier contoured seats and interiors, including door cards and wind-down windows as opposed to the sliding-section window.

The Wolf's most noticeable difference to previous Land Rovers was the repositioning of the spare wheel onto its flank. Given the weight of the tyre and wheel set, a special sling was developed to enable safe retrieval. The interior was totally redesigned and the more comfortable civilian-style seats with

Series III towing arrangements: a fine shot of the NATO-standard tow hitch and electric power socket. (Mark Griffiths)

Wolf TUL interior: looking rearward, this shot shows the internal roll cage, commercial-style seats and plastic door card. (Ben Skipper)

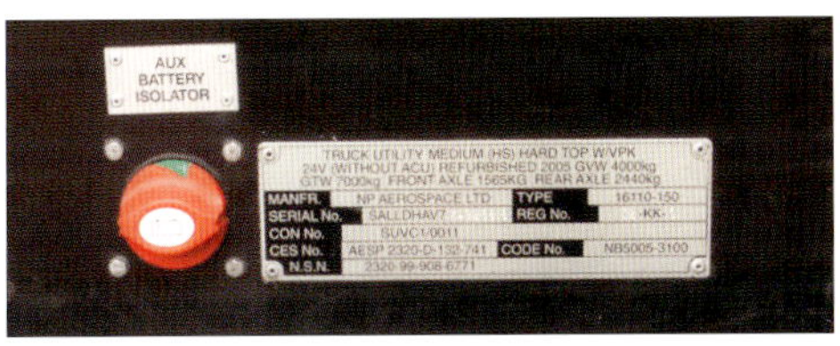

CAV 100 interior: the battery isolation switch underneath the commander's seat. (Ben Skipper)

Series III LWB: Matt Black and Bronze Green paint finishes soon fade in direct light to a 60 percent grey tone and light green. Note the plastic grille and tactical bridge plate. (Mark Griffiths)

Left: Defender 90 interior: note the radio-mounting rack behind the seats and clips for holding crew weapons; there's also a rather neat locker between the seats for maps, snacks and bits and pieces. (Mark Griffiths)

Wolf TUL tie-down hooks fitted to the tailgate and sides are essential for keeping the tilt on. Note the white face of the rear differential for reflecting convoy light. (Ben Skipper)

the headrests were added, as well as a rudimentary plastic door card.

The map light was now a flexible stalk instead of being fitted to the dashboard on the passenger side. The wing-mounted air vent, which was omitted from the original Defender range, reappeared, this time fitted to both sides allowing the snorkel to be fitted according to local conditions.

CAV 100 Land Rovers operating in desert environments were often fitted with a heat-reflective cover attached by Velcro and strapping. Internally there was a range of unique features including a Bowden cable used to raise the external mesh windscreen protector. A battery isolation switch was installed beneath the driver's seat, and the heavy front doors were fitted with extra catches to prevent opening from the outside.

A large air blower and bulkhead light were fitted between and slightly behind the driver and passenger for comfort. The troop-carrying compartment featured a roof observation panel that was covered by a forward-sliding panel and surrounded on all four sides by thick Perspex shielding, which could be folded away from troops. Internally, a fold-down curtain was installed between the front and rear cabs to prevent leaks of light, debris and to afford visual protection from elevated points. The heavy rear doors could be locked by the commander via remote control to prevent external opening.

Specifications

Due to the many variants of the Land Rover, I have selected the specifications of key models that would be of interest to modellers.

Series I: Truck, ¼ Ton, Cargo, General Service 4x4; FV18001; Rover Mks 1–5
Production period: 1948–1958
Body/chassis: Welded box-section steel ladder chassis with aluminium panelled body over steel frame
Engine: Rover 4 cylinder-in-line
Cubic capacity: 1,595 (1.6l)–1,997cc (2.0l) petrol, 2,052cc (2.0l) diesel
Horsepower: 1.6l petrol: 50bhp at 4,000rpm; 2.0l petrol: 52bhp at 4,000rpm; 2.0l diesel: 51bhp at 4,000rpm
Torque: 1.6l petrol: 80lbf/ft at 2,000rpm; 2.0l petrol: 101lbf/ft at 1,500rpm; 2.0l diesel: 87lbf/ft at 2,000rpm.
Carburettor: Petrol models: Solex
Transmission: 4F1Rx2; part-time 4x4 (earlier models were full-time 4x4 with freewheel)
Steering and suspension: Recirculating ball, or worm and nut. Live axles on multi-leaf semi-elliptical springs; hydraulic double-acting telescopic shock absorbers
Brakes: Hydraulic drums system throughout
Electrical systems: 12v
Dimensions (80in model):
Track: 50in (1.27m)
Wheelbase: 80in (2.03m)
Total length: 132in (3.35m)
Overall width: 62in (1.58m)
Height: soft top: 70–76in (1.78-1.93m); hard top: 72in (1.81m)
Dry weight: 2,721lb (1,237kg)

Series IIA, Series III 'Lightweight': Truck, ½ Ton, General Service, Fitted For Radio (FFR) 24v, 4x4; FV18102; Rover 1
Production period: 1965–1984
Body/chassis: Welded box-section steel ladder chassis with aluminium demountable body panels over steel frame
Engine: Rover 4 cylinder-in-line
Cubic capacity: 2,286cc (2.25l) petrol; 2,286cc (2.25l) diesel
Horsepower: 2.25 petrol: 70bhp at 4,000rpm; 2.25 diesel: 62bhp at 4,000rpm
Torque: 2.25 petrol: 124lbf/ft at 2,500rpm; 2.25 diesel: 103lbf/ft at 1,800rpm
Carburettor: Zenith 36 IV
Transmission: 4F1Rx2; part-time 4x4
Steering and suspension: Recirculating ball, or worm and nut; option of drag link mounted steering damper. Live axles on multi-leaf semi-elliptical springs; hydraulic double-acting telescopic shock absorbers.
Brakes: Hydraulic drums system throughout. Series III models have vacuum servo assistance
Electrical Systems: 12v or 24v
Dimensions:
Track: 52in (1.31m)
Wheelbase: 88in (2.24m)
Total length: assembled: 147in (3.73m); stripped: 143in (3.63m)
Overall width: assembled: 64in (1.63m); stripped: 60in (1.52m)
Height: assembled: 77in (1.96m); stripped: 58in (1.47m)
Dry weight: 3210lb (1,459kg)

Series III LWB: Truck, ¾ Ton, Cargo, General Service 4x4; FV18071; Rover Series III
Production Period: 1971–1985
Body/chassis: Welded box-section steel ladder chassis with aluminium panelled body over steel frame
Engine: Rover 4 cylinder-in-line
Cubic capacity: 2,286cc (2.25l) petrol; 2,286cc (2.25l) diesel
Horsepower: 2.25 petrol: 70bhp at 4,000rpm; 2.25 diesel: 62bhp at 4,000rpm
Torque: 2.25 petrol: 124lbf/ft at 2,500rpm; 2.25 diesel:103lbf/ft at 1,800rpm
Carburettor: Zenith 36 IV
Numerous engine options were offered including Rover/Buick V8
Transmission: 4F1Rx2; part-time 4x4
Steering and suspension: Recirculating ball, or worm and nut; option of drag link mounted steering damper. Live axles on multi-leaf semi-elliptical springs; hydraulic double-acting telescopic shock absorbers. Aeon rubber helper springs were offered as options
Brakes: Hydraulic drums system throughout; optional vacuum servo assistance
Electrical system: 12v or 24v
Dimensions:
Track: 52in (1.31m)
Wheelbase: 109in (2.77m)
Total length: 142in (3.61m)
Overall width: 66in (1.68m)
Height: 77in (1.96m)
Dry weight: 3,301–3,455lb (1,500–1,570kg)

Forward-Control 101: Truck, 1 Tonne, General Service, Forward Control; FV19000 series; Land Rover
Production period: 1972–1978
Body/chassis: Welded box-section steel chassis with aluminium demountable body panels over steel frame
Engine: Rover/Buick V8
Cubic capacity: 3,528cc (3.5l) petrol
Horsepower: 91–114bhp at 3,500rpm
Torque: 166–185lbf/ft at 2,000rpm
Carburettor: Twin SU
Transmission: 4F1Rx2; full-time 4x4; positive lock-on inter-axle differential
Steering and suspension: Recirculating ball, or worm and nut; Woodhead steering damper. Live axles on multi-leaf semi-elliptical springs; front anti-roll bar; double- acting telescopic hydraulic shock absorbers.
Brakes: Drum system throughout; servo-assisted dual circuit hydraulic
Electrical system: 12v or 24v

Dimensions:
Track: 61in (1.55m)
Wheelbase: 101in (2.57m)
Total length: 162in (4.12m)
Overall width: 72in (1.83m)
Height: soft top: 84in (2.13m); hard top: 94in (2.39m)
Dry weight: 4,233lb (1,591kg)

Defender 110: Truck, Utility, Medium (TUM), 4x4; Land Rover Defender 110
Production period: 1983–2016
Body/chassis: Reinforced welded box-section steel chassis with aluminium panelled body over steel frame
Engine: Rover 4 cylinder-in-line
Cubic capacity: 2,495cc (2.5l) petrol; 2,495cc (2.5l) diesel
Horsepower: 2.5l petrol: 83bhp at 4,200rpm; 2.5l diesel: 68bhp at 4,200rpm
Torque: 2.5l petrol: 133lb/ft at 2,000rpm; 2.5l diesel: 114lbf/ft at 1,800rpm
Carburettor: Weber 32/34 DMTL
Numerous engine options were offered including Rover/Buick V8 and Td4
Transmission: 5F1Rx2; Full-time 4x4; centre lockable differential
Steering and suspension: Recirculating ball, or worm and nut; power assistance optional. Live axles on long-travel, dual-rate coil springs; front axle location by Panhard rod with rear 'A' frame locator; hydraulic double-acting telescopic shock absorbers
Brakes: Disc throughout; duel servo-assisted hydraulic
Electrical system: 12v or 24v
Dimensions:
Track: 67in (1.70m)
Wheelbase: 110in (2.794m)
Total length: 175in (4.45m)
Overall width: 70in (1.78m)
Height: 80in (2.03m)
Dry weight: 4,717lb (2,144kg)

Land Rover Variants

No other military support vehicle has been modified as extensively as the Land Rover, though even a design as flexible as the Land Rover's couldn't fulfil some tasks. Various design teams have risen to the challenges thrown at them by the various procurement agencies to produce some truly unique vehicles. Sometimes the ideas were simply not practicable for a range of reasons, but most of the time the designers and engineers got it so right that they created legends. The Land Rover bore a huge range of variants, and here follows those available to modellers as kits or conversions.

Carmichael Redwing FT/6. This wonderful fire engine is presented in the colours of the Army Fire Service, fully equipped and ready to go. (Alf Van Beem)

Carmichael FT/6 Redwing Fire Appliance

The Carmichael Redwing was a Series IIA LWB conversion used by the Army Fire Service in their larger storage depots, where its low profile was extremely useful for movement between and into buildings. Finished in red, the FT/6 was a Forward-Control type vehicle based on the 109in chassis to which a front extension was added. This meant the driver and front passenger were relocated forward of the original bulkhead that now acted as a strengthening element, with the remaining crew sitting behind.

The body was a mix of modified and new panels which covered hose reels, water tanks and pump assemblies. Delivery hoses were carried in roller-shuttered side lockers. The FT/6 was also equipped with a sliding ladder which extended to 35ft. It was fitted with a Coventry-Climax pump which when in operation could deliver up to 1,600 litres of water per minute, in addition to the 635 litres carried on board. An oil cooler was fitted to allow prolonged pumping as the pump was driven from the gearbox power take-off. Suction hosing was also carried to enable water to be drawn from an emergency water supply, or open source.

TAC-T and TACR

The Truck, Fire-Fighting, Airfield Crash – Tactical, or TAC-T appeared as a result of the RAF's increasingly field-based role with the growing Harrier force entering service in 1972. A formidable piece of field kit, it provided advanced firefighting and rescue options. However, it was soon withdrawn from the role as it was unable to handle the new Aqueous Film Forming Foam (AFFF) that was introduced into service in mid-1970s.

The Series II-based Truck Aircraft Crash Rescue-1 (TACR) replaced the ACRT from 1970 onwards in the first-responder role for all three services. Supplied as a truck cab mounted on a special commercial one-ton chassis, the TACR was fitted with a purpose-built body by HCB-Angus. This featured a 100-gallon water tank, a high-pressure pump with side outlets, 100ft of hosing, dry-powder extinguishers and an oscillating light.

The cab featured a hatch for elevated firefighting; a powerful telescopic Plessey floodlight and a Francis searchlight were also fitted. An extendible ladder was carried; an outward-facing seat for a third crewmember was installed to the rear. These vehicles were finished in either Fire Engine Red or Bronze Green.

Showing off its Series III heritage, this TACR is full of detail; note the water tank gauges on the pillar behind the passenger's door. (Alf Van Beem)

This former RAF Carmichael TACR2 rescue crash tender, complete with ancillaries, is now in service with the Yorkshire Air Museum as an airport crash tender. (Brain Burnell)

The Carmichael TACR2 finished in tactical drab; note the hessian sheeting above the windows, ready for rolling down to prevent sun glare. (Allen Watkin)

The TACR2, made by Carmichael of Gloucester, was predominantly used by the RAF and based on a heavily modified Range Rover. This 6x4 vehicle was designed as a fast-response vehicle with deliveries starting in 1977. It carried four firefighters had an on-board tank with a 900-litre capacity, which could deliver foam though a series of easily accessible foam branches.

The main stowage locker was accessed from the rear, behind a sliding shutter, while a folding ladder, light equipment and oscillating light were fitted on top. TACRs were supplied in Fire Engine Red, Khaki Drab or Royal Blue for Queen's Flight units.

Pink Panther

The Pink Panther LWB Land Rover was the first custom-made vehicle for SAS use. Previous attempts to produce a suitable vehicle had highlighted engine and chassis issues. FVRDE stepped in and identified the technical and mechanical requirements that needed to be met. Marshalls of Cambridge was tasked with making the prototype and by mid-1968 had produced the first vehicle. This was given heavy-duty springs and shock absorbers along with a steering damper.

The chassis was strengthened at its critical points and guards added to protect key drive train components. Sills, doors, windscreen and tilt were removed and the spare tyre was fitted to the front bumper to allow clearance for the bulkhead-mounted GPMG.

Along with additional storage, sand channels and pioneer tools were placed along the outsides of the cargo bed which housed the extended-range fuel tanks that afforded a range of some 1,500 miles. A drop-down section was mounted on the rear on which further stores could be carried.

The cargo area was home to ammunition and weapon stowage, a rear-facing GPMG mount, radio and rear crewmember position. Metal rifle containers were mounted on the wings and electronically fired smoke-grenade launchers were placed on the bumpers either side of the spare wheel. As well as personal equipment, radios, a sun compass, a standard magnetic compass and a theodolite were also carried. All this equipment added almost 500kg to the overall weight of the vehicle, which handled the weight well and the platform proved itself to be reliable.

WMIK

The most recognizable of contemporary British Land Rovers, the WMIK, pronounced 'Wimmik', standing for Weapons Mount Installation Kit, was a joint venture between Ricardo Special Vehicles and Land Rover. This turned the Wolf into a formidable fire support vehicle. Fitting started in 1999, with the kit made up of a modular roll cage and weapons mounts specifically designed for Wolf Defenders. As a result, the WMIK could carry an array of support weaponry including the MILAN ATGM, or anti-tank guided missile. Doors and windows were removed and a snorkel fitted.

In 2006 the MoD upgraded a series of former CAV 100 Wolf Defenders to HP (high payload) WMIK standard, which became known as Enhanced or E-WMIK models, while the original vehicles were known as O-WMIK. The E-WMIK was able to carry the new Heckler & Koch automatic light grenade launcher (ALGL), Manroy heavy machine gun or Javelin missile launcher. Bowman radio sets were also carried as well as ECM equipment to counter mobile-phone signals. It used the CAV 100 wheels with their run-flat inserts, along with increased personal protection, including underfloor mine armour. To cope with the extra weight, the chassis, brakes and suspension were upgraded.

The final version of the WMIK was the R or Re-Life vehicle. This had an extra crew position, upgraded electronics, chassis and suspension. The engine was enlarged to 2.8l and a new automatic gearbox fitted. Other changes included drivers getting their own roll cages, slab armour to protect crew as well as IED detectors. A small cargo carrier was fitted to the bonnet for cam-net or light baggage carriage; a new flat-fronted grille was also fitted. These later R-WMIK had fifteen-spoke wheels. A foul-weather kit of canvas tilt and side doors was also produced.

Series IIA LWB Marshalls of Cambridge SAS Land Rover. Designed in conjunction with the SAS, this Land Rover would earn the nickname 'Pink Panther' for its unique desert camouflage. (Alf Van Beem)

A WMIK of the RAF Regiment on patrol at Kandahar Airfield, Afghanistan. (Efren Lopez, USAF)

This flank view affords the modeller a clear view of the unique profile of the R-WMIK that's normally hidden under patrol stores. (Peter Harvey)

Scout

The Scout is a special stripped-down reconnaissance vehicle which has had its doors, windscreen and side windows removed. These vehicles are fitted with a roll cage which runs front to rear, over the cargo bay area. Similar in appearance to the WMIK, it lacks the mounted weapon points and bonnet cargo carrier. A tarpaulin is often fixed over the driver and passenger compartment to provide rudimentary weather protection, while the front vents are covered with a waterproof shroud. The Scout is in service mainly with Yeomanry regiments.

130 XD

This version of the Wolf family is mainly used as a Pulse Battlefield Ambulance (BFA) and is fitted by Marshalls of Cambridge. Powered by the same diesel engine at the TUL and TUM, the 130 XD BFAs are available as air-conditioned desertized, waterproofed or winterized versions. Pick-up truck versions are also in use with mortar platoons.

Composite Armoured Vehicle (CAV) 100 Snatch/Vixen

The CAV 100 replaced the Series VPK Land Rovers in Northern Ireland from 1992 onwards. Designed as a new vehicle based around the Truck Utility Medium (TUM) chassis, it was fitted with light composite ballistic armour known as

An action shot taken of a BFA 130 preparing to move a casualty at Lashkar Gah, Afghanistan, 2009. (POA(Phot) Dave, MoD)

CAMAC. Made of glass fibre and plastic, this armour was originally produced by Courtaulds Aerospace. As the armour was built to high tolerances, the Land Rover military engineering team designed a new 110in chassis to accommodate the new bodypack and front cabin armour panels. The vehicles were then assembled by GKN who were experienced in working with armoured vehicles.

Defender 130: a great study in camouflage at the Hari Merdeka Celebrations, Kota Kinabalu, Sabah, Malaysia, 2013. (Uwe Aranas)

The ultimate Snatch CAV 100 model, the well-appointed Vixen complete with thermal insulation. (Cpl Russ Nolan, RLC, MoD)

The vehicle's official designation is CAV 100, with 1,300 units produced by NP Aerospace for internal security use by the British Army. Known as Snatch because of its role in being able to take troops safely through crowds to enable them to rapidly apprehend participants in civil disturbances, the vehicle was armoured against small-arms fire and shrapnel.

Snatch-1s made up the initial 1,000 units delivered and fitted with the V8 petrol engine. Snatch-1.5 was the 24v model intended for desert use. Snatch-2 was an LHD training model featuring 12v electrics while Snatch-2A was a 24v RHD version fitted with air-conditioning packs in the storage lockers behind the driving compartment. These vehicles were fitted with a 3.5l petrol engine generating 114bhp. The CAV 100 was soon upgraded to take the 300Tdi engine which produced 111bhp; these upgraded vehicles became known as the Snatch-2B.

A small number were upgraded to Vixen configuration by automotive engineer company Ricardo. This upgrade applied the lessons learned in Afghanistan and Iraq and featured a strengthened chassis, a modified front axle, a completely new rear axle and modified suspension to cope with increased payload. The project also included an upgraded engine, new automatic transmission and increased front cabin belly protection.

Although withdrawn from use by the British, the CAV 100 remains in use globally, protecting diplomatic assets as well as third-sector and news personnel in high-tension areas.

Desert Patrol Vehicle

This coil-sprung Marshalls of Cambridge vehicle replaced the previous leaf-sprung vehicles in the late 1980s. Like its predecessor, the Desert Patrol Vehicle (DPV) was a stripped model, this time fitted with the cargo bay found on civilian Land Rovers. Virtually identical in stores carriage, the DPV was fitted with an external roll cage behind the driver and passenger seats.

RSOV

Built by Land Rover's Special Operations Vehicle department, the RSOV or Ranger Special Operations Vehicle is a Defender

Left: The SAS Defender 110 Desert Patrol Vehicle shows off its commercial heritage. (Alf Van Beem)

110-based weapons platform which has been in use with the US Rangers since 1992. Purchased to replace their Jeep-based M151 Light Gun Trucks, the RSOVs fulfil three major roles: weapons carrier, communications vehicle and medical vehicle.

The weapons carrier is able to carry up 8,000lb of stores as well as seven Rangers. It is armed with several machine guns and either a .50cal (12.7mm) heavy machine gun or the fearsome Mk 19 grenade launcher. A mortar carrier, known as a MORTSOV, is also available for mortar platoons at battalion level. The communications vehicle carries the necessary equipment to facilitate reliable communications between platoon, battalion and task force HQs while in the field. The final variant, the MEDSOV, is equipped with six stretchers and, as well as driver and commander, can also carry up to three medics. Looking very similar to the Wolf WIMK, the RSOV is finished in matt black and features a low-profile roll cage, with storage cages along the sides and back of the cargo bay.

Below: Fitted out in a medical evacuation role, this US Ranger RSOV is on display at the National War College, Washington DC, USA, 2001. (Michael Lemke)

JRA Perentie 6x6

The Jaguar-Rover-Australia Perentie is the name given to Australian-produced Land Rovers. The Perentie is based on the 110 chassis, is powered by an Isuzu 3.9l four-cylinder diesel engine and is available as either the 4x4 or 6x6 vehicle.

Fulfilling all the traditional roles of the Land Rover from cargo carrier to FFR, the 6x6 can also operate in the medical, air defence and special forces role. In the latter, this unique-looking Land Rover is capable of performing long-range patrols and often carries a 250cc motorcycle on its rear flank.

Perentie 6x6. Showing off a two-tone finish at the 2013 Australian Defence Force Academy open day, this 6x6 is ready for patrol. (Nick Dowling)

In Service & In Action

Right: A US Federal Government-commandeered Series I Land Rover serving with the US Army Engineers. Pyongyang, North Korea, 1950. (US Department of Defense)

Far right: A stripped Australian Army Series I Land Rover negotiating a ditch in South Vietnam. (Australian War Memorial)

The Land Rover was never intended to be a fighting vehicle; it was purchased to fulfil liaison and administrative tasks such as transporting stores, yet within months of introduction Land Rovers were supporting British troops in Korea. Its baptism of fire and ensuing reputation put it in the motor transport pools around the world for the next seventy years. It would be produced under licence by companies such as Minerva, Santana and Otokar; it would also be supplied in kit form to countries such as Brazil, New Zealand and Zaire.

On parade at Kajman, Plzeň, Czech Republic, this Defender 130 SOV exudes firepower, 2015.

This Czech Army WMIK SOV pauses while on manoeuvres at Brdy, Czech Republic. (Stanislav Rada)

The Land Rover found itself deployed wherever there was a British presence. As the empire shrank and new states were formed, Land Rovers were often left behind for the new owners. Many former colonial states experienced some form of conflict during independence and Land Rovers found themselves increasingly in the front lines.

In Malaya their speed and versatility were key to containing and combating the communist resistance as the British soon realized that meeting the insurgency with large military formations was not going to work. Here the Land Rover supported troop and police actions in a low-key manner that some contemporary armoured fighting vehicles could not.

In Aden mine-proofed Land Rovers with their large role circular roll cage supported ground actions against the indigenous insurgencies; they were ideally suited to the mountainous terrain and narrow streets of the main centres of population. It was also used by the RAF in the mountain rescue ambulance role to support operations from RAF Khormaksar.

By the late 1960s, the Land Rover was in use with the Australian Army working with the Americans in South Vietnam and with the SAS in Oman during the Dhofar Rebellion. SAS use saw the Land Rover embrace the role of special forces support vehicle. Painted pink to help it merge in with the Omani terrain, especially in low-light conditions, the Pink Panthers

An RUC 'Greenfinch' poses with a well-worn Series III VPK Land Rover. (Ronnie Gamble)

supported the SAS as they reconnoitred and countered the insurgents. The SAS also operated their 'Pinkies' in Northern Ireland though these retained their original Bronze Green finish.

The Pink Panthers remained in SAS until the mid-1980s when replaced by Defender-based vehicles. These would be used against Iraqi targets to devastating effect during the first and second Gulf wars as well as undertaking reconnaissance. Such was their success that the US Army Rangers and others adopted the Land Rover as their replacement Special Operations Vehicle.

By the 1970s the Land Rover could be found supporting military units around the globe. It was also during this period that the Land Rover developed a new role as an armoured personnel carrier. The Land Rover-based Shorland provided excellent protection for the Royal Ulster Constabulary (RUC) and Ulster Defence Regiment (UDR) in their border patrols between Northern Ireland and Eire.

Its aggressive appearance restricted its use in urban areas, and with Series Land Rovers deployed by the Army to counter increasing IEDs and small-arms fire, the introduction of the VPK was a step forward. It would be used extensively for security patrols in the province during the 1970s and 1980s. The VPK was replaced by the CAV 100 Snatch in the early 1990s, which was well suited to the role, but was soon outclassed on operations in Afghanistan and Iraq.

During the Cold War Land Rovers continued to support military operations throughout the world. With the thawing of the Cold War and the collapse of former communist states, the Land Rover was deployed once again, finished in the white of the UN peacekeeping forces. The Balkans suited the Land Rover, the roads often narrow and in poor state of repair, leaving heavier vehicles unable to operate. As the situation deteriorated and NATO became involved, Land Rovers returned to their camouflage finish. These would be adorned with an array of signage including IFOR (Implementation Force), SFOR (Stabilization Force) and Kosovo Force (KFOR) in large white lettering.

Mk 1 Shorland armoured car.

Aircrew prepare to extend the nose ramp of a C-5B Galaxy aircraft to offload British Land Rovers at Grootfontein logistics base, Namibia, 1990, to be used by Finnish UN troops (MSgt Bill Thompson, USAF)

CAV 100s of the British Army; part of the Organization for Security and Co-operation in Europe, Kosovo, November 1998. (Bela Szandelszky/OSCE)

CAV 100s were also used by British personnel operating for the Organization for Security and Co-operation in Europe (OSCE) as part of the Kosovo Verification Mission. Their task was to ensure that Yugoslav forces were abiding by the UN October 1998 ceasefire agreement. These

Right: TUM Medical Emergency Response Team (MERT) ambulance, Camp Bastion, Afghanistan, 2011. (Sergeant Alison Baskerville, RLC, MoD)

Below: Perentie 6x6 LRPVs saw service with Australian Special Operations Task Groups in Afghanistan, 2009. (Leading Seaman Paul Berry, 1st Joint Public Affairs Unit, Australian Defence Force)

vehicles were painted orange and retained their military number plates. Some were fitted with a rectangular blue OSCE decal on the vehicle sides behind the driver and passenger doors.

In 2000 the Parachute Regiment and their pathfinders would deploy to Sierra Leone as part of Operation Palliser. Their WMIK Wolfs were used to project force and support foothold operations, as well as to reassure civilians and show numerous militia gangs that the British meant business.

In 2002 Land Rovers were deployed as part of Operation Herrick (Afghanistan) and Operation Telic (Iraq). TULs, TUMS, WMIKs and CAV 100s would be deployed in both theatres. The CAV 100s had served the Army well in the internal security role, especially in Northern Ireland and Kosovo, but in the new asymmetrical theatres of Operation Herrick and Telic they were of little use. The use of massive IEDs by insurgents that could knock out AFVs took a heavy toll on the CAVs. As a result of public and political pressure, they were withdrawn from 2006 onward and replaced with more suitable vehicles such as the Mastiff. WMIKs continued to serve, primarily providing fire support and on convoy protection duties as well as fulfilling reconnaissance roles.

The Australian SAS initially used their 6x6 Perenties as long-range patrol vehicles (LRPVs) to patrol the interior of Australia. In Afghanistan and Iraq, the LRPVs were used as platforms to support reconnaissance and deliberate attack missions. When the LRPV left theatre in 2013 they were transferred to the Australian 1 and 2 Commando Regiments.

Today the Land Rover continues to serve the British military as the R-WMIK, as well as TUL and TUM versions. Overseas users, including the Italian Carabinieri and US Rangers, also continue to use their Land Rovers to fulfil operational requirements.

Airmobile asset: a Defender of the Latvian Special Operation Forces loads up at Rukla, Lithuania, 2010. Note the bonnet basket and sand tracks. (Isaac A. Graham, US Army)

Up and away: TUM and trailer being airlifted by RAF Chinook, RAF Waddington Air Display, 2010. (Dan Davison)